ATLAS OF KENTUCKY

ATLAS OF KENTUCKY

CONTRIBUTORS

Wilford A. Bladen

Thomas P. Field

Terry L. McIntosh

Phillip D. Phillips

Karl B. Raitz

Richard I. Towber

William A. Withington

P. P. Karan and Cotton Mather, Editors

David Oakes, *Cartographer*

John Fraser Hart and James E. Queen, *Consultants*

Gyula Pauer, *Map Design*

THE UNIVERSITY PRESS OF KENTUCKY

Photo Credits

Frontispiece: Louisville, Kentucky's largest city. Approximately one-fifth of the state's population lives in the Louisville metropolitan area. Located at the Falls of the Ohio, it was chartered on May 1, 1780, by Thomas Jefferson, then governor of Virginia. The city is first in the state in industry and trade as well as population.

ISBN: 0-8131-1348-2

Library of Congress Catalog Card Number: 76-24337

Copyright © 1977 by The University Press of Kentucky

A statewide cooperative scholarly publishing agency serving Berea College, Centre College of Kentucky, Eastern Kentucky University, The Filson Club, Georgetown College, Kentucky Historical Society, Kentucky State University, Morehead State University, Murray State University, Northern Kentucky University, Transylvania University, University of Kentucky, University of Louisville, and Western Kentucky University.

Editorial and Sales Offices: Lexington, Kentucky 40506

The position of photographs is indicated by the following abbreviations: B (bottom), C (center), L (left), R (right), T (top).

Albert B. Chandler Medical Center, Lexington, 53, 55, 57
Ashland Daily Independent, 87
Dudley Bentle, 26 (R), 35, 56, 62, 95 (L), 132, 149
Berea College, Office of Information, 49 (R)
Wilford A. Bladen, 65 (R)
Darbie Bowers, WKYT-TV, Lexington, 69 (BR)
Brown and Williamson Tobacco Corp., Louisville, 89 (TL)
Fred J. Burkhard, Liberty, 145
Calumet Farm, Lexington, 109, 141 (R)
Jimmy Carter Campaign Headquarters, Lexington, 172 (TR)
Courier-Journal and *Times*, Louisville, 10 (L), 11 (C), 31 (BR), 70 (R), 108, 117
Billy Davis III, ii
Charles Doyle, 65 (L), 70 (BL), 73, 105 (R)
Eastern Kentucky University, 26 (BL)
Filson Club, Louisville, 13 (C)
Firestone Textiles Company, Bowling Green, 89 (BR)
Herald-Leader, Lexington, 29 (L), 70 (TL), 76 (L), 76 (R), 92, 172 (L), 173
Lonnie W. Hodges, University of Kentucky College of Agriculture, 138
P. P. Karan, 64 (R), 125 (R)
Keeneland Association, Lexington, 11 (L), 141 (L)
Kentucky Department of Agriculture, 121 (L), 129, 130, 131, 137, 146
Kentucky Department of Education, 46, 54
Kentucky Department of Military Affairs, 171
Kentucky Department of Natural Resources and Environmental Protection, 160
Kentucky Department of Public Information, 13 (L), 13 (TR), 13 (BR), 17 (R), 22, 26 (TL), 29 (BR), 36 (TL), 36 (BL), 45, 64 (L), 78 (BL), 96 (TL), 96 (BL), 96 (R), 105 (L), 114 (L), 119 (T), 119 (B), 120 (T), 121 (L), 121 (R), 133, 143 (L), 144, 151 (L), 152, 153, 154, 155, 168, 169 (T), 169 (B)
Kentucky Development Cabinet, Information and Communication Section, 14, 163 (L)
Kentucky Division of Forestry, 123 (L), 123 (R)
Kentucky Division of Photogrammetry, 21 (L), 21 (R)
Kentucky Educational Television, 69 (TR), 69 (L)
Kentucky Kernel, Lexington, 29 (TR), 172 (BR)
Kentucky New Era, Don Smith, xii
Lexington-Fayette Urban County Police Division, 36 (TR), 36 (BR)
Cotton Mather, 10 (C), 10 (R), 11 (R), 17 (L), 18 (R), 20, 31 (TL), 33 (L), 33 (R), 40, 44, 52, 58, 89 (TR), 95 (BR), 127, 142 (L), 142 (R), 143 (R), 147, 151 (TR), 164
Murray State University, Information and Public Services, 49 (L)
National Distillers and Chemical Corp., New York, 91 (L), 91 (R)
National-Southwire Aluminum Co., Hawesville, 86
J. Noye, Big Sink Farm, 139
Karl B. Raitz, 18 (L), 24, 31 (BL), 31 (TR), 32, 60, 67, 78 (TL), 79, 95 (TR), 104, 114 (R), 118, 120 (B), 125 (L), 126, 136, 148, 151 (BR), 163 (R), 170
U.S. Department of the Interior, 78 (R)

Contents

Foreword

When Daniel Boone and other early settlers came to Kentucky, they were impressed by the richness and variety of the landscape. Today Kentucky is growing and changing but the variety remains. The natural regions of the Bluegrass State have long been a training ground and field laboratory for scholars, bringing many of the leading geographers of the United States to Kentucky for fieldwork and training.

We Kentuckians are justly proud of our state. We are eager to display the beauty of our natural landscapes and the friendliness, enthusiasm, and accomplishments of our people. It is fitting that we should have an atlas to aid in the portrayal of these features. This ATLAS OF KENTUCKY will be invaluable, not only to Kentuckians, but also to scholars, teachers, planners, business leaders and government officials everywhere. My heartiest congratulations to Professors P. P. Karan, Cotton Mather, and other contributors on an outstanding cartographic reference source.

Julian M. Carroll
Governor of the Commonwealth of Kentucky

Preface

THE MAKING OF A THEMATIC ATLAS

Although the idea of producing an atlas containing thematic maps is far from new, the history of modern comprehensive thematic atlases of individual American states covers less than a quarter of a century. In the United States such volumes have usually been prepared by academic institutions rather than by the official cartographic services or agencies, which have concentrated their efforts on topographical and special-purpose mapping.

American state atlases vary considerably in many respects. The size and shape of a state and the nature of its environment, economy, and population necessitate differences in map design and format. The "ethos" of a state also reveals itself in some ways in the choice of map topics. Further, the increasing use of state atlases for planning purposes influences the selection of map topics.

The *Atlas of Kentucky* provides a concise and up-to-date view of the Commonwealth's physical, human, and economic patterns. As a reference document it is a graphic representation of the various phases of Kentucky's geography and development. Subjects have been selected for coverage on the basis of their usefulness in conveying an understanding of significant social, economic, physical, and resource management questions. The maps on various themes compress, abstract, and present data in a form that clarifies the spatial relationships and juxtapositions relevant for purposes of regional planning. Such basic spatial quantities as distance, direction, connectedness, contiguity, and areal extent—essential information inputs in comprehensive planning—may be analyzed through study of the various thematic maps. By transforming complex patterns into simpler patterns, the maps also serve as information filters and facilitate comparison of spatial distributions and relationships between two or more phenomena. Maps on topics such as health, residential preference, environmental quality, and urban structure, which are commonly not found in state atlases, have been included here because of their importance in social and environmental planning.

The creation of an atlas takes considerable planning, researching, and designing in order to portray as much information as possible on the maps within the limitations of book size and format. Subjects for map coverage were chosen through consultations among the editors and contributors, then were assigned to individual contributors. Multiple authorship has marked advantages for an atlas of this scope. First, each contributor was a specialist on his topic. Second, each concentrated his attention on a specific topic, thus enabling intensive research on the subject. The task of the general editors was to check and revise maps and text materials from all contributors and consultants to achieve a unity in the overall pattern.

Because the sound graphic portrayal of material through the language of maps depends greatly on design considerations, a considerable amount of time was devoted to this vital step. A scale must be chosen for each map—a most important factor in presentation of information. The format of this atlas was determined by the standard one-to-a-page map. The dimension of this standard map derives from the shape and size of Kentucky. The scale chosen is large enough to show basic distributions while at the same time permitting the book to be used comfortably at a desk. For the representation of highly detailed phenomena the standard map was enlarged to a double-page format or was reduced in scale for simple choropleth and isarithmic maps. By varying the scales we have been able to choose map formats suitable to the complexity or importance of various subjects, to amplify some subjects, and to show the more interesting phenomena in detail. The use of successive scales also simplifies the comparison of different maps and enables three or four maps with secondary scales to be placed in the format of a major map. In the format and cartographic design of the atlas we have tried to strike a balance between content and visual cognition so as to enhance the appearance and clarity of the maps.

The value of maps in a state atlas depends in part on the particular symbols and methods of depiction chosen, on comprehensive content within the limits of map scales, and on judicious generalization. While it is not necessary to belabor these considerations, some explanations are needed regarding our choice of methods of depiction. In making this choice we have taken into consideration, first, the peculiarities of the geographic distribution of the phenomena being mapped, and second, the cartographic need to find the particular technique that will most effectively portray those phenomena.

Graded symbols such as circles or cubes have generally been used here to depict magnitudes of distribution of areal phenomena. Isarithm or lines joining points of equal value have been used to show the distribution of selected physical and behavioral elements. These lines indicate the gradients, troughs, rises, and other configurational characteristics of the phenomena. The technique known as choropleth mapping, in which enumeration units such as counties are shaded to indicate values within each unit, has been employed to represent phenomena with discrete areal distribution. The choice of interval in distribution obviously affects the precision of a choropleth map, for the details of the distribution which lie within the interval are buried; the larger the interval, the more detail will be lost. The intervals we selected are small enough so that the lost information is kept to a minimum, and the sizes of the intervals are consistent with the quality and quantity of data. We have tried to select the most appropriate interval for each map out of a number of possibilities.

Symbols in modern cartography have come a long way from the first maps scratched in the sand by early man. Where the map maker of the past used simple symbols and relied primarily on his eyes, his art, and his legend, the modern cartographer has a wide variety of symbols to assist him in creating an effective means of visual communication. We have, for the most part, used widely accepted standard symbols and colors. By the use of graduated screens, the four standard colors (black, yellow, red, and blue) have been combined to produce a wide variety of color tints chosen to be aesthetically pleasing as well as scientifically functional.

The choice of appropriate type styles and sizes is important to the clarity, legibility, and visual impact of a map, and also determines the character of the map. In general we have used Optima Semibold for map titles, Optima for legend, Century Expanded for cultural features, and Century Expanded italics for hydrologic features. We have tried to avoid visual trauma by limiting the number of type styles, sizes, and degrees of boldness, and differences in placement. Further, only those names, symbols, and lines are included which are essential to orient the map reader or which have some significant relationship to the concept being communicated. County boundaries made in recessive pastel colors by the

use of screens facilitate areal identification without loss of focus on map patterns, and also avoid distracting clutter. The transparent overlays with county names in the back pocket of this atlas are designed to help the reader locate specific counties.

Extensive research was necessary to acquire, evaluate, and verify the information required for the preparation of high quality new maps. The contributors to this atlas carefully selected sources of information, using the most recent and accurate available. In many cases personal contacts with local, state, and federal agencies have enabled us to obtain data that have not yet been published. In some cases data were collected through field survey and questionnaires. Readers interested in sources of information should consult the section at the back of the atlas.

The next step was to prepare original map manuscripts. These hand-drawn compilations showed different classifications of information over appropriate base maps, within the approved layout and graphic design. The map compilations were then thoroughly reviewed and revised where necessary.

Although the various processes in the preparation of the maps may be clearly distinguished, they must in practice be considered in conjunction from the outset. Each part of the production demands coordination, skill, and experience if the maps are to be accurate as well as aesthetically pleasing. After design, research, and compilation of manuscript maps, three steps were involved in the preparation of final maps: drafting, color proofing, and preparation of final four-color composite film negatives for the printer. Cartographic technicians skilled in drafting, type and symbol application, and tone separation, prepared the map drawings from the original compilations. Linework was scribed by hand, using precision instruments with interchangeable points that engrave uniform lines on a translucent, dimensionally stable coated film. Lines were scribed at reproduction scale in color-separated form. Each linework element formed an original drawing and reproducible negative. All elements were pin registered to assure accurate positioning. Symbols and type were positioned on overlays, which were contacted to make negatives.

The map linework elements were used to etch the emulsion of peel-coat films to prepare negatives for each color tone. The emulsion was removed from areas inside the etched lines, creating "windows," which were then exposed through screens of various densities to produce the desired values of each color. The various map elements—linework, type, and peel-coats—were then combined to produce a composite negative for each color.

Photomechanical and chemical processing of the composite negatives to make color proofs was the next step in preparation of the maps. Color proofs were used to verify color values and design and to check all linework, symbols, and type. Four-color composite film negatives of map elements, reflecting final decisions on color, were prepared for the printer as the last step in the cartographic preparation of this atlas. A bibliography of selected references on cartographic methods and techniques used in this atlas is given at the back of the book.

ACKNOWLEDGMENTS

The editors wish to thank all those who contributed their knowledge and skill to the making of this very complex book. The seven contributors listed on the title page are all members of the Department of Geography at the University of Kentucky. In addition to these contributors, Harry H. Bailey, Richard Booth, Ronald Garst, James Newman, John Ragland, Milton Shuffet, Geoffrey Wall, Stephen E. White, M. D. Whitaker, and Allan Worms participated in map compilations or review of certain materials. Valuable help was also received from C. Boerner, Brenda Oakes, I. Shair, and Cynthia and Bill Wright.

Virtually all cartographic work was done by David Oakes of the United States Geological Survey in Lexington, Kentucky. Gyula Pauer of the University of Kentucky Department of Geography did all the photowork.

Throughout the preparation of the atlas valuable comments and criticisms were received from many geographers and cartographers across the country. The atlas has gained especially from the editorial comments of John Fraser Hart, past editor of the *Annals* of the Association of American Geographers; and from the counsel of James E. Queen, chief cartographer in the illustrations section of the United States Geological Survey. The late Professor Arch Gerlach, a geographer at the United States Geological Survey and chairman of the National Atlas Committee, provided valuable design suggestions.

The editors express their utmost appreciation to a native Kentuckian, Professor W. A. Bladen, who, in addition to being a contributor, assumed responsibility for verifying numerous facts and figures to ensure an accurate and authoritative portrayal of contemporary patterns in Kentucky. They are also deeply indebted to various agencies of the state government in Frankfort for providing data used in the compilation of many maps. These agencies are identified in the listing of data sources.

Significant from the outset of the project was the encouragement of Dr. Lewis W. Cochran, Vice-President for Academic Affairs at the University of Kentucky. Dr. Cochran's support and assistance were vital in the initiation and evolution of the atlas.

Introduction

Kentucky is a regional tapestry whose physical strands are interwoven with the variegated threads of human experience. Its richness rests on the variety of its physical elements: relief, rocks and minerals, soils, climate, and animal and plant life. This is a land that has nurtured persons of spirit and commitment. The pulse of its people has been represented by such men as Daniel Boone, Henry Clay, Jefferson Davis, Abraham Lincoln, John James Audubon, Irvin S. Cobb, and John Jacob Niles. With courage, statesmanship, the painter's brush, humor, and song, these men led in social and artistic exploration and development, expressing in their lives the character of the land of Kentucky.

In an age when the men of America assumed that this land was their kingdom alone, Kentucky produced women of fine and noble spirit—women such as Carry Nation, Mary Breckinridge, Josephine McGill, and Alice Hegan Rice. It was on this land, too, that Ellen Churchill Semple was born. A geographic scholar who attained national recognition as president of the Association of American Geographers in 1921 and who won international prestige with her *American History and Its Geographic Conditions*, she was to influence a generation of scholars in the social sciences. Some of her finest work was centered in the Kentucky mountains.

Carl O. Sauer, the founder of cultural geography in America, and the internationally recognized geographer Preston E. James, together trained graduate students at Mill Springs, Kentucky, during the 1920s and 1930s. In that same period Willard R. Jillson, state geologist of Kentucky and a student of Rollin D. Salisbury, who founded the first graduate department of geography in America at the University of Chicago, commissioned such distinguished geographers as Darrell H. Davis, John B. Leighly, Carl O. Sauer, and Kenneth C. McMurry to write a series of monographs on the state's regions. This admirable series constitutes one of the most intensive and exhaustive studies ever made in the United States of the people, activities, environment, and resources of an individual state.

Kentuckians have a strong sense of place and region, and their speech mirrors their perception. Their appreciation of place is exemplified by names such as Solitude, Society Hill, Golden Eagle Cliff, Windy Gap, Polecat Hollow, Back Slough, Gap of the Ridge, Tater Hill, Dry Fork, Lickskillet, Cedar Bluff, and Hell For Certain. The regional names also indicate continuity of place reference and are carried over into the new structures that are changing the face of Kentucky, as in the Cumberland, Mountain, Blue Grass, Purchase, and Green River parkways. The importance of place names to Kentuckians is manifested in many other ways—putting county names on automobile license plates and introducing persons with reference to the region in which they live. In short, regionalism is a pervasive and salient feature of Kentucky.

Regional pride is a highly developed characteristic of Kentuckians. There is pride in the regional flora: the dogwood, redbud, sourwood, maple, willow, and sycamore, the trillium, bloodroot, lady's-slipper, columbine, and blue aster. There is pride in such regional foods as corn pudding, country ham, and fried chicken. Pride is evidenced in the structures placed upon the land. Indeed, few states have such notable public buildings, such imposing homes, or such fine reconstructions of historic sites. Pride is expressed, too, in the local community, including the courthouse square with its traditional landscaping and its historic plaques. In short, perhaps more than any other American state, Kentucky has real regional style, reflecting local traditions, customs, and geography.

With the vivid geographic character of the Kentucky land, with the heritage of the Commonwealth as a training ground for geographers of national note, and with a rare series of regional monographs produced here a half century ago, it now seems appropriate to present this atlas of the physical, economic, and cultural patterns of the land of Kentucky.

Conceptually, the scope of this atlas of Kentucky is broad. It recognizes that geography is essentially a study of the character of a place. "Place" is both an area and an aggregation of points; it has both external and internal relationships, it has a time dimension, it is both physical and human, it is economic and social, and it has political relevance. Place is of interest to all who have any curiosity, but it is of central significance to those who are concerned with a planned and ordered application of physical and human resources for the betterment of our environment.

The rapid growth and spread of population in Kentucky, together with the necessity to protect the environment, have led to a need for more spatial information about the various regions of the state. The thematic maps in this atlas are intended to fill that need. Their principal function is to reveal the patterns of distribution of a wide range of environmental elements, resources, economic activities, and demography. A reader can, for example, take any region of Kentucky and explore many of its essential characteristics by comparing selected thematic maps. This book should provide an essential information source for policy makers, regional planners, and many others.

The atlas should also serve to stimulate interest and pride on the part of many Kentuckians in the cultural, scenic, and economic diversity of the Commonwealth. Behind current newspaper headlines are factors of geography, physical characteristics, population, economic resources, cultural heritage, and political attitudes that have exerted and will continue to exert a strong influence on the destiny of the Commonwealth. Those who have an understanding of the spatial patterns of these factors can relate current developments to the physical and cultural environment of Kentucky.

For the general reader it should be pointed out that an atlas such as this differs from the usual book in a fundamental way. Rather than using words as the primary means for communicating ideas, theories, and relationships, an atlas composed of maps uses symbols, colors, and patterns. It provides a large amount of spatial information in coded form, and for this reason the maps must be carefully examined and interpreted before their significance can be fully understood. They can communicate only if the reader takes the trouble to develop skill in reading the various types of maps and other visual materials included.

The texts appearing in previous state atlases vary greatly in their aim, content, and length. In this atlas the text consists of an explanatory description of the spatial pattern shown on a map or on a related group of maps. For better understanding of the subject of the maps, the text is supplemented by photographs and occasionally by tables and diagrams. The text is also intended to prolong the usefulness of the maps, especially the economic maps. Despite changes in data, distribution patterns often remain essentially unchanged over a period of time. Maps based on data from the 1970 census have been updated by the inclusion of more recent data in the text whenever the pattern has remained stable. Where the pattern has changed since 1970 the maps have been entirely reworked. The texts are placed as near the maps as possible to facilitate simultaneous use.

The maps are organized according to several themes developed in consultation with colleagues at the University of Kentucky and elsewhere. The relationship of these themes to the overall scope and purpose of the atlas is described below.

THE LAND

The introductory maps showing the hypsometry and physical framework (**Map 1**) and the land regions (**Map 2**) are designed as reference points to facilitate the use of the thematic maps that follow. The relief, drainage network, and general physical character of Kentucky revealed on **Map 1** help to identify and put into perspective the geographic patterns of various socioeconomic and cultural phenomena depicted on other maps.

The map of Kentucky's land regions (**Map 2**), delineating areas with a high level of homogeneity, is of interest not only for its geographic information but also for its practical value in planning. Regional diversities and peculiarities of the physical environment should be taken into account in planning the state's economy and in implementing measures designed to bring major changes in the landscape.

EXPLORATION AND EARLY SETTLEMENT

The patterns of exploration and early settlement of Kentucky (**Map 3**) provide an essential background for the understanding of contemporary patterns.

POPULATION CHARACTERISTICS

The economy, culture, and political life of a state are manifestations of the social activity of its population. Several characteristics of Kentucky's population have therefore been selected for cartographic representation (**Maps 4–16**). This group of maps shows the distribution and growth of population, age structure, migration, residential preference patterns, distribution of the black population, and church affiliations.

Maps of socioeconomic levels (**Map 17**), distribution of income (**Maps 18 and 19**), crime (**Maps 20 and 21**), distribution of the work force in various economic activities (**Maps 22–25**), and housing characteristics (**Maps 26–29**) should prove useful in dealing with the problems of development in the Commonwealth.

Maps of educational levels (**Maps 30–32**) and health services (**Maps 33–37**) are of interest in developing the cultural and physical aspects of Kentucky's human resources. Most existing state atlases have given insufficient attention to these vital aspects influencing the quality of human resources. Maps of population and of the quality of human resources, since they deal with the principal productive force in society, are the most important and therefore precede those on the state's economic geography.

TRANSPORTATION AND COMMUNICATION

Maps 38–42 show the extent, importance, and capacity of the state's transportation system. The map of highway traffic flow indicates the extent to which various highways are used. The maps of railroad routes, navigable waterways, airports, and bus routes show economic ties between different areas of Kentucky and between the state's major production centers. Communication maps such as those of newspaper circulation (**Maps 43–46**) and households with television sets (**Map 47**) are rare in state atlases.

URBAN MAPS

Although the bulk of the population of the United States lives and works in urban areas, the evolution of urban systems, the commuting patterns, and the structure of urban agglomerations have not received adequate treatment in existing state atlases. The chapter representing aspects of the urban system (**Maps 48–61**) should facilitate the understanding of Kentucky's urban evolution, functional relationships between major urban places and the surrounding countryside, and the distribution in major urban areas of selected socioeconomic factors such as age of housing, income, and black population. By presenting the urban realities in graphic form these maps should prove useful in city planning.

MANUFACTURING AND TRADE

The map of Kentucky's manufacturing (**Map 62**) shows the geographic distribution and importance of industrial centers. **Map 63** reveals that Kentucky's industrial subsystem is firmly entrenched in the nationwide entrepreneurial system, as shown by the corpo-

rate linkages with other states. **Maps 64 and 65** show the pattern of two industries for which Kentucky is well known, bourbon and tobacco.

The maps of retail sales and wholesale trade (**Maps 66 and 67**) show the spatial structure of Kentucky's business and commercial aggregations. The map of retail sales indicates that variations in the volume of retail sales are a function of regional variations in population and income characteristics. The map of the wholesaling system reveals that the location of distribution centers depends on the degree of market concentration, the intensity of demand, and the adequacy and efficiency of the communications network.

GEOLOGY, MINERALS, AND ENERGY

The geological maps (**68 and 69**) provide information on the structure and development of Kentucky's earth crust and give a general picture of possible locations of mineral wealth. They also offer insights into the relationship and influence of structure and lithology on the peculiarities of the state's soil (**Map 77**) and land use potential (**Map 75**). Tectonic maps are generally not found in state atlases, being relatively recent; one has been included here because of its value in understanding geologic structures and the distribution of useful minerals. The geology map has been supplemented with cross-sections showing the salient formations and tectonic movements.

The map of mineral resources (**Map 70**) is of great practical use. Kentucky's major mineral resource, coal, is the subject of two maps (**71 and 72**). The map of the electric power industry (**Map 73**) shows all the important generating stations in the state and indicates their capacity and type of plant. **Map 74** deals with the extraction, processing, and transportation of oil and gas.

LAND USE AND PHYSICAL ENVIRONMENT

The maps showing physiographic regions, regional land use, soil associated with physical areas, and use-suitability of land areas (**Maps 75–78**) are an important reference work for overall agricultural planning and development, for soil improvement programs, for application of agricultural technology, and for the study of other practical problems. The climatological maps (**79–88**) indicate areal distribution of various elements which are relevant to Kentucky's agricultural production, such as moisture, temperature, and dates of first and last frosts (indicating the length of growing seasons). Some of the maps, such as average annual days with measurable precipitation (Map 81), also have practical value for location of outdoor recreational sites. Similarly, the maps with dates of the first and last frosts are valuable in the building and construction trades.

FORESTRY AND AGRICULTURE

Map 89 is designed to give a clear and detailed picture of the geographic distribution of commercial forests in Kentucky. The importance of forests to both the physical environment and Kentucky's economy makes this map very useful.

Farming occupies an important position in Kentucky's economy. The agricultural maps (90–114) show the land in farms, harvested cropland, distribution of farms according to class, distribution of selected major crops, land in pasture, distribution of different types of livestock, farm income, and levels of mechanization.

RECREATION

Maps 115–119 show the location of historic sites, monuments, and scenic areas which are objects of pride within Kentucky. For out-of-state visitors they should serve to stimulate an interest in the Commonwealth's history, culture, scenic parks, and tourist centers. Of special interest are Maps 116 and 117, showing the numbers of overnight guests and visitors to parks in the state. The economic input of these visitors has a major impact on Kentucky.

AIR AND WATER QUALITY

Clean air and water are important components of environmental quality. A map of the ambient air quality (Map 120), generally absent from other state atlases, has been included to give an overall idea of the relatively clean air which prevails over most of Kentucky. The map of river basins (Map 121) depicts the patterns of surface waters. The map is useful for estimating water resources available for various purposes. The map of streams affected by mine drainage (Map 122) is of value in planning for the protection of stream water quality. Map 123, showing towns and cities in the state which face water supply problems, should be of practical value to local and state agencies.

POLITICAL AND ADMINISTRATIVE STRUCTURE

The state's administrative and territorial structure is shown on Maps 124–134. Those maps showing the evolution of the state's principal administrative units, the counties (Maps 124–127), are especially interesting for the view they provide of the growth of the Commonwealth. The contemporary county map serves as a base for all choropleth maps in this atlas. The administrative structure of such functions as regional or area development, the judiciary, elections, and liquor sales does not always coincide with county administrative divisions, and maps have been included to show the separate territorial divisions of these services or institutions. The map of military bases and national guard units in the state reflects their importance in Kentucky's economy.

VOTING PATTERNS

Maps 135–142 show the areas of influence of the two main political parties and the growth or lessening of the patterns of this influence during the 1967–1975 period. The results of gubernatorial and presidential elections have been used to indicate the regional dominance of the political parties in various areas of Kentucky.

The establishment of a harmonious relationship between society and nature within the context of rational use and conservation of resources is the ultimate goal of the people of Kentucky. The maps in this atlas, particularly those depicting physical conditions, land use, population and health factors, economic activities, water and air quality, and recreational resources, should provide a valuable resource for the study of the influence of man on his environment. It is our hope that through analysis of these maps the people of Kentucky will be enabled to trace past tendencies, to reflect upon the changes now taking place, and to plan for the future protection of the land of Kentucky.

P. P. KARAN and COTTON MATHER

ATLAS OF KENTUCKY

PHYSICAL FRAMEWORK

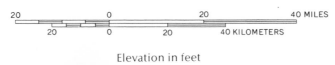

Elevation in feet

3001 and above
2001–3000
1001–2000
401–1000
0–400

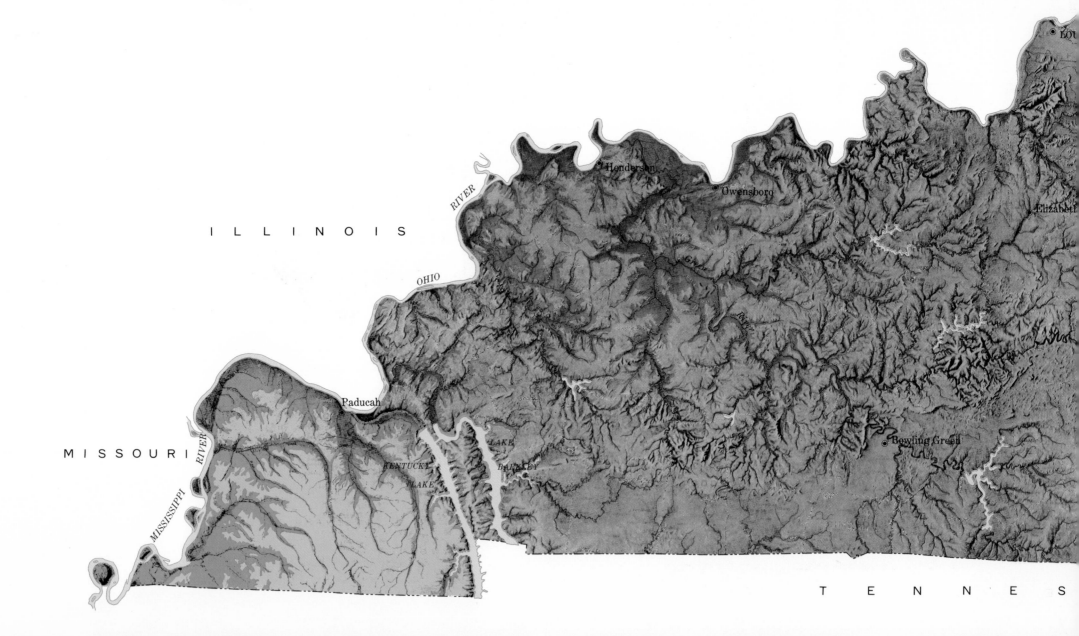

INDIANA

LOU

Henderson

Owensboro

Elizabeth

ILLINOIS

RIVER

OHIO

Paducah

MISSOURI

RIVER

LAKE

KENTUCKY
LAKE

LAKE
BARKLEY

Bowling Green

MISSISSIPPI

TENNES

MAP 1

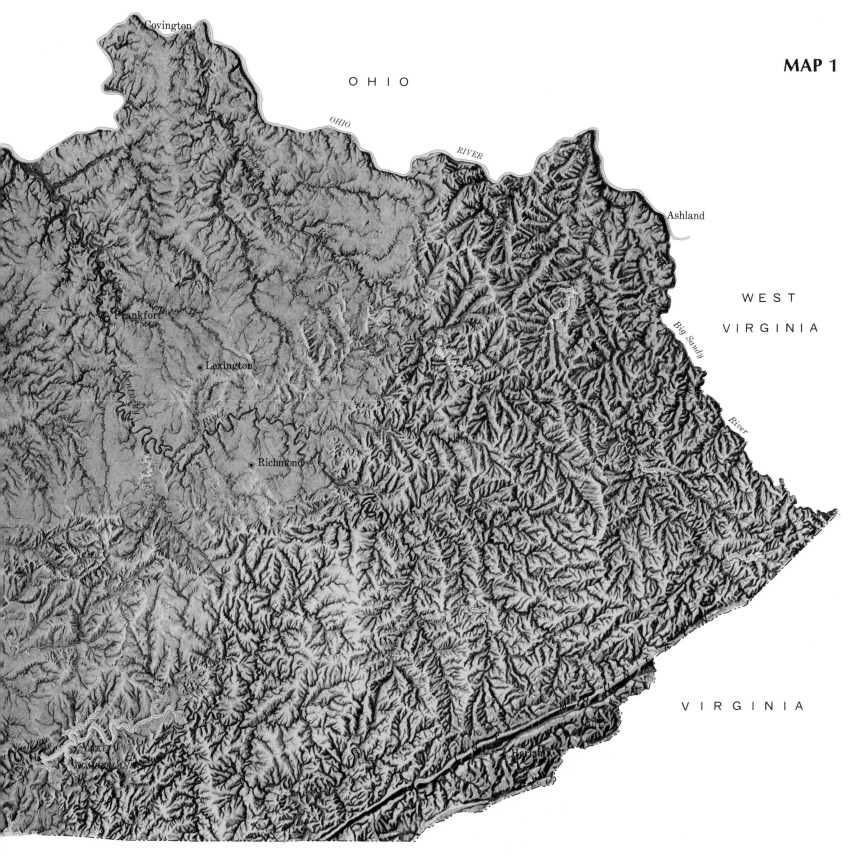

OHIO

OHIO

RIVER

Ashland

WEST

VIRGINIA

Big Sandy

River

Covington

Frankfort

Lexington

Richmond

Kentucky

River

VIRGINIA

Harlan

E

I. THE LAND OF KENTUCKY

MAPS 1 and 2

The land of Kentucky is the stage on which have developed the physical and human characteristics portrayed in this atlas. It is characterized by a variety of terrain, from the Ohio and Mississippi River floodplains through the rolling uplands to the Appalachians, and ranges in elevation from approximately 400 to over 4,000 feet above sea level. This varied terrain is associated with diverse physical and cultural patterns. The physical landscape and the social and economic characteristics of the Jackson Purchase in western Kentucky, for example, stand in sharp contrast to those of the Mountains of eastern Kentucky. Subsequent maps in this atlas will be more meaningful if the reader bears in mind the physical characteristics of the land shown on **Map 1** (previous pages).

The northern boundary of Kentucky was established in 1784 when Virginia (of which Kentucky was then a part) gave up claim to the land north of the Ohio River. The low-water mark on the north shore of the Ohio was agreed upon as Kentucky's northern boundary. An exception is Green Island, opposite Henderson, where a portion of land north of the Ohio is part of Kentucky. The southern boundary, formed by the extension of the Virginia-North Carolina line in 1780, then called the Walker Line, was supposed to follow the 36° 30' parallel from the Virginia line to the Tennessee River. Because of an error in surveying, however, the boundary was placed north of that parallel. The error was never corrected and the boundary alignment as it now appears was made official in 1820. When the Jackson Purchase was completed in 1818, the Ohio River boundary on the north was extended to the Mississippi River and the 36° 30' parallel was accepted as the southern boundary from the Tennessee River to the Mississippi.

The eastern boundary begins where the Big Sandy River joins the Ohio and follows the main northeasterly branch of the Big Sandy to Pine Mountain, then follows the ridge of Pine and Cumberland mountains to the Tennessee line. In 1799–1800 the eastern boundary was surveyed and demarcated on the ground. Kentucky's western boundary follows the middle of the Mississippi River and includes river islands 1, 2, 3, 4, 5, and 8. Number 5 island is Wolf Island. An interesting feature of the western boundary is New Madrid Bend of the Mississippi River, an exclave of Kentucky completely cut off from the rest of the state.

Kentucky is comprised of five distinctive regions (**Map 2**): the Jackson Purchase, the Western Coal Field, the Bluegrass, the Mountains,* and the Pennyroyal.

The Jackson Purchase, in the westernmost part of the state, is culturally a part of the Deep South. For years after the purchase of this area in 1818 from the Chickasaw Indians, the region was effectively isolated to the east by the Tennessee and Cumberland rivers. Along the north and west are the Ohio and Mississippi rivers; to the south only the state boundary marks the region apart from Tennessee. Physically, the Jackson Purchase ranges from the fertile alluvial lowlands of the Mississippi, through hills and the Barrens (broad natural grasslands), to the Breaks of the Tennessee. Development has centered on Paducah, a great port on the Ohio and a railway center astride the rail line from Chicago to New Orleans. The city, which has superhighway connections, handles tobacco, soybeans, and livestock from the region's productive

*The term Mountains is used here in preference to Cumberland Plateau, another frequent name for this region. "Mountains" refers to the dissected nature of the topography; "plateau" refers to the geologic structure. For similar reasons Pennyroyal is used here in preference to Mississippian Plateau.

LAND REGIONS

MAP 2

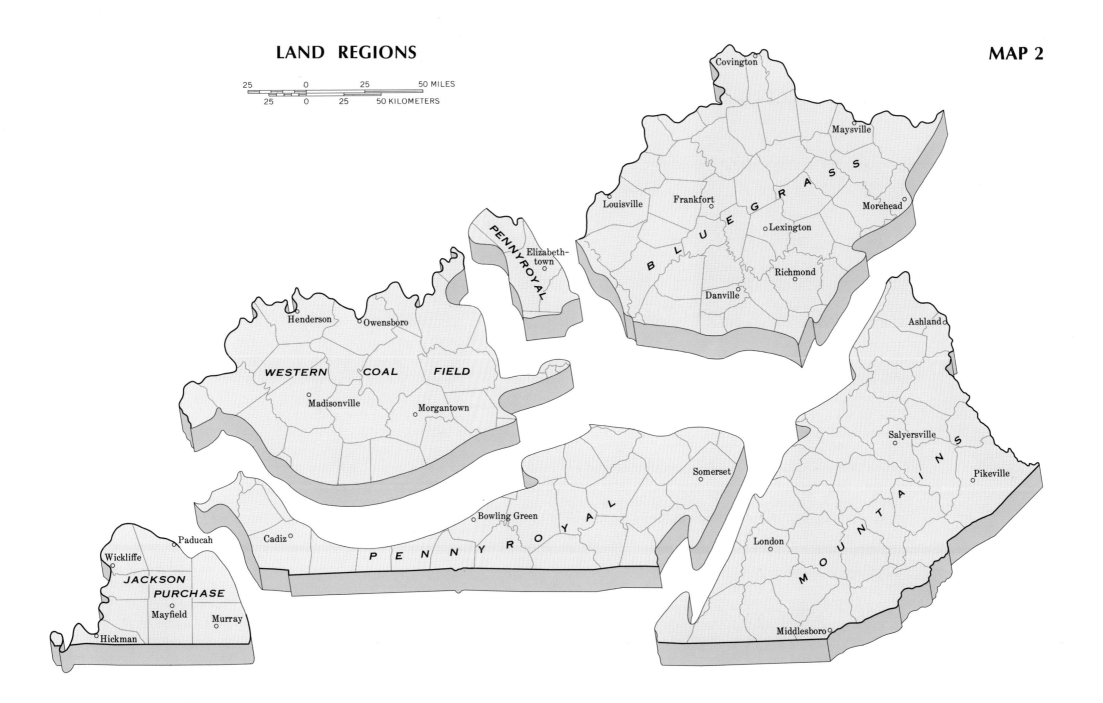

Above left: Alluvial lowlands along Madrid Bend of the Mississippi River, at the extreme western end of the Jackson Purchase.

Center: Massive mechanized coal-mining operations in the Western Coal Field Region, where about half of Kentucky's coal is produced.

Right: A view of the Knobs, rounded hills of Devonian and Silurian rocks which surround the Bluegrass Region.

farms. Paducah is also the focus of industrial expansion and power generation along this part of the Ohio River. Significant power development and a booming tourist industry are associated with Lake Barkley and Kentucky Lake, formed by the damming of the Cumberland and Tennessee rivers, and with Land Between the Lakes recreational area.

The Western Coal Field, an oval region bordered on the north by the Ohio River, has energy resources of vital national significance. The region varies from hilly coal-mining country in the south to floodplain along the Ohio River. Underground, strip, and auger methods of coal extraction are used, although strip mining is most important. As coal production has soared, man has become an awesome geomorphological agent, and environmental problems (stream pollution and soil erosion) have become regional issues. Manufacturing, the chief source of regional employment, is located mostly around the three major urban centers, Owensboro, Henderson, and Madisonville. Agriculture is modestly developed in the southern counties, but the farmers to the north in Union, Webster, Henderson, Daviess, and McLean counties are important producers of corn, soybeans, and hogs.

The Bluegrass Region, with Lexington near its center, includes Louisville at its western margin. The Louisville area, the state's largest metropolitan development, has approximately one-fifth of the total population of Kentucky. The city's location at the Falls of the Ohio gave it an early impetus as a strategic transportation site. Now the metropolitan district is a great rail, highway, water, and air transport center and the main and most diversified industrial area in the state.

The Bluegrass Region is renowned for its distilleries, its spectacular horse farms, and its productive beef and tobacco enterprises. Both the Inner and the Outer Bluegrass are underlain by limestone whose partial removal by ground water has produced a gently-rolling karst plain. The parklike landscape of the gentleman farms near Lexington is internationally famous. The breeding, training, and racing of Thoroughbred and Standardbred horses have spawned numerous subsidiary activities in the region, including recreation on which tourists expend over seventy-five million dollars annually. Lexington, a booming urban center and the home of the University of Kentucky, is the world's largest burley tobacco market and an important industrial and medical center. Surrounding Lexington are the charming county seat towns of Georgetown, Paris, Winchester, Richmond, Nicholasville, and Versailles. Frankfort, the state capital, is in the midst of bourbon country. Kentucky has twenty-seven distilleries; all but four are in the Bluegrass.

The Mountains Region, one of the state's two largest regions, representing about one-fourth of the total area, includes Pine and

Cumberland mountains in the east. Elevations range up to 4,150 feet in this extremely rugged land. While the most significant and long-range endowment of the region is its natural scenery, the present economy is focused upon coal. Kentucky is one of the nation's leading coal producers and slightly over half of the state's production is from this region. Somewhat more coal is gotten from underground than from surface operations. Other major economic activities are lumbering and marginal agriculture. Most settlements are on the narrow floodplains and lower mountain slopes, while the railways and highways parallel the streams.

The Mountains are a cultural as well as a natural region. Distinctive ideas and traditions characterize the region's society. Problems of isolation, poverty, education, and health plague many communities. While program planning for the region has been long range, little has been done about implementation. Because of the rugged terrain, large areas remain essentially unsettled and the average population density for the region is below the state average. The average density on the limited favorable terrain is high, however, and the landscape manifests many regional problems.

The Pennyroyal is the state's most extensive region. It reaches from the Bluegrass and the Mountains to the Jackson Purchase. In a great arc the Pennyroyal nearly encircles the Western Coal Field, reaching the Ohio River near both Paducah and Louisville. In a very

real sense the Pennyroyal is not a centrality but a region between places. It stretches from the rolling Bluegrass to the smooth lands of the Jackson Purchase, from the coal field of the eastern part of the state to the Western Coal Field. It lies between the metropolitan centers of Louisville and Nashville and between eastern and western Kentucky with their specialized interests in agriculture, industry, tourism, and services.

Physically, the Pennyroyal is limestone country—relatively high and considerably dissected in the east, a gently rolling plain in the central part, and boldly demarcated by sandstone escarpments in the Clifty area (the internationally famous cave country of Kentucky).

Although the Pennyroyal contains no large city, most of the people are urban oriented. There are numerous modest-sized urban centers in which aggregately many people live. The countryside is also the home of many commuters who work in those urban areas. The region's chief center for industry, farm commodities, wholesaling and retailing, transportation, service, and education is Bowling Green. Numerous important urban and industrial functions of the Pennyroyal are shared by Hopkinsville, Elizabethtown, and the smaller cities. Although a diversified agriculture prevails over most of the region, the two mainstays of the prosperous farming system are tobacco and livestock.

COTTON MATHER

Above left: Internationally famous Calumet Farm, producer of several Derby winners, near Lexington in the Inner Bluegrass.

Center: The rugged terrain of the Mountains of eastern Kentucky, an important coal-mining and lumbering region.

Right: A tobacco barn on the gently rolling plain of the central Pennyroyal, an area noted for its agricultural production.

II. EXPLORATION AND EARLY SETTLEMENT

MAP 3

At all levels of political, economic, and social development, the creation of a territorial communications network depends upon topography, perceived need, and freedom of access. For the many centuries of Indian occupation of Kentucky we can perhaps never know how these three forces interacted to produce the communication pattern that prevailed at the time of pioneer European settlement. We do know that the land now occupied by the state of Kentucky presents little in the way of serious barriers to travel by foot and small boat.

In the latter half of the eighteenth century freedom of access in Kentucky was achieved through territorial penetrations of most of the Indian tribal lands by the "league of the Iroquois" incursion from New York westward to the Mississippi River and southward to the Tennessee River. The Cherokee, linguistic cousins of the Iroquois, together with their allies, the Catawba, drove the Shawnee out, but the Cherokee never occupied the land abandoned by the Shawnee, "the dark and bloody ground." Thus the door was open for Indian trails, explorer routes, and later pioneer roads leading to settlements.

The succession of Indian tribes across the broad area east of the Mississippi had communication routes based on trade, contact with related tribes, and warpaths between enemy tribes. It is possible, too, that some of the well-known trails led to seasonal hunting grounds and to the vital salt deposits.

Some trails followed easy topographic routes while others made use of protective cover for covert movement. Most of the trails il-

lustrated are only fragments of trade routes and warpaths which led from the Great Lakes country into the Deep South. It was these trails in combination with the relatively strong barrier of the Appalachian highlands to the east that created a strong north-south bias to the trail pattern of Kentucky.

The motivation for European exploration of Kentucky was both economic and political. The French, coming up the Mississippi River, explored major rivers, marked their mouths with lead plaques, and assessed the land for furs and minerals. The British explorers came across from the New River country of Virginia or down the Ohio to assess the potentials of the land for settlement. Both French and British explorers used access routes established by the Indians. (It would be most helpful if we knew something of Indian explorers!) Only for surveying purposes or through fear of attack would an explorer blaze a new trail through unfamiliar country when the easy way of an established Indian trail was available.

Map 3 through color and boldness of line, emphasizes the land routes and downgrades the water routes of the state. The Ohio and Tennessee, however, were major arteries for both the Indians and the explorers. The trails were created simply to give access to places away from the rivers. Such explorers as Christopher Gist, Thomas Walker, and Daniel Boone used only land routes. The pioneer coming to settle the land already explored, James Harrod for example, made maximum use of the rivers to transport things which were required for settlement.

The Indian trails—footpaths—are difficult for a population on wheels to perceive. When the way is made so easy by the cut and fill of the bulldozer it is hard to remember the over and down and through of the trail. The descriptive fragments that have come down through history are difficult to relate to, even for those who have backpacked along the Appalachian Trail. The most common destination of the modern trail, the top of the mountain, was almost unknown for Indian trails. Such a destination would offer little to the traveling Indian.

These trails, so clearly defined on the map, were really "flow lines" rather than single well-defined roads. The trail would be in one place in wet weather, in another in dry periods. And in peacetime certain routes would be used that were unsafe during war. Topographic features such as Cumberland Gap would constrain the traveler's use of alternate routes even in times of stress. Once the trail was free of topographic restraints, however, a series of essentially parallel trails and traces could be used at the traveler's discretion.

To the modern traveler in Kentucky this map of the Indian and explorer trails provides the essential clues for visiting a number of historic places (**Map 115**). Many otherwise ordinary places and scenes become most interesting when their history is known. Usually the state guide to historical markers serves to pinpoint the key locations.

THOMAS P. FIELD

The lookout at Cumberland Gap National Historical Park (*left*) offers a view of three states. Through this passage streamed over 100,000 settlers between 1775 and 1800. Daniel Boone (*above center*) was instrumental in opening Kentucky to settlement through Cumberland Gap, as was James Harrod, who led settlers down the Ohio River. Fort Harrod, built in 1775, has been reconstructed on its original site (*above right*). General George Rogers Clark, one of the leaders in securing northern Kentucky against Indian harassment and later a hero of the Revolution, is memorialized in a statue at Louisville (*above left*) overlooking the Ohio River.

THE INDIAN AND EXPLORER TRAILS

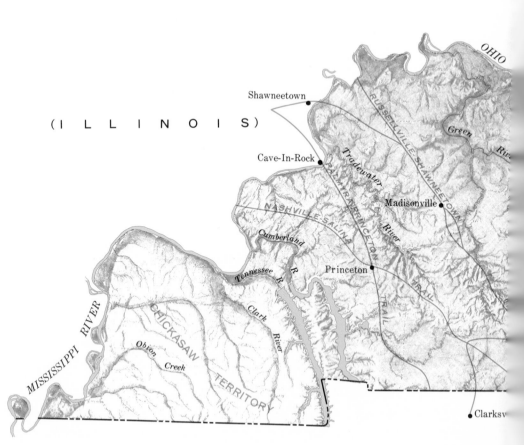

(ILLINOIS)

Shawneetown

Cave-In-Rock

Madisonville

NASHVILLE-SALINA

Cumberland

Tennessee R.

Princeton

Clark River

MISSISSIPPI RIVER

CHICKASAW

Obion Creek

TERRITORY

OHIO

Green River

Tradewater River

RUSSELLVILLE-SHAWNEETOWN

SALINA-PRINCETON TRAIL

Clarksv

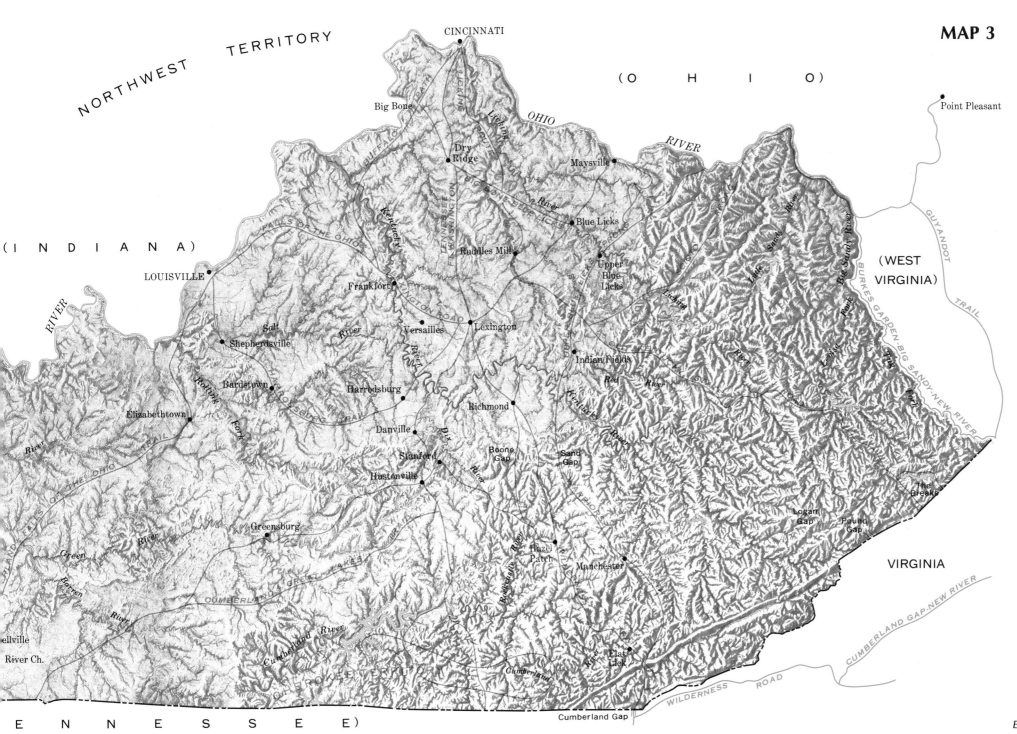

MAP 3

NORTHWEST   TERRITORY

(O   H   I   O)

• Point Pleasant

Big Bone

OHIO

(INDIANA)

RIVER

Dry
Ridge

Maysville

Blue Licks

Licking

(WEST
VIRGINIA)

FALLS OF THE OHIO

LOUISVILLE

Ruddles Mills

Upper
Blue
Licks

Frankfort

RIVER

Salt

River

Versailles

Lexington

Indian Fields

Kentucky

Red

River

Little

GUYANDOT

TRAIL

Licking

Big Sandy River

Sandy River

BURKES GARDEN-BIG SANDY-NEW RIVER

Levisa

Tug Fork

Shepherdsville

Bardstown

Harrodsburg

River

Richmond

Kentucky

River

Dix

Boone
Gap

Sand
Gap

Logan
Gap

Pound
Gap

The
Breaks

Elizabethtown

River

Rolling

Fork

Danville

Stanford

Hustonville

River

Greensburg

River

Green

Barren

River

ellville

River Ch.

Cumberland River

Rockcastle River

Hazel
Patch

Manchester

VIRGINIA

River

Flat
Lick

Cumberland

River

CUMBERLAND GAP-NEW RIVER

WILDERNESS   ROAD

T E N N E S S E E)

Cumberland Gap

CINCINNATI

III. POPULATION CHARACTERISTICS

MAPS 4-7

DISTRIBUTION, DENSITY, AND GROWTH

Kentucky's 1970 population, as enumerated by the United States Census, was 3,218,706, an increase of 5.9 percent over the 1960 population of 3,038,156. The state's most rapid percentage growth in population occurred between 1790 and 1820 **(Table 1)**, when pioneer settlers were streaming across the Appalachians, filling the valleys of the eastern Mountains, the Bluegrass, and other areas which were attractive to agriculturists. Since 1820, Kentucky has grown more slowly than the United States as a whole in every decade except the 1930s. In 1820 Kentucky represented 5.9 percent of the national population; by 1970 this had fallen to 1.6 percent. The growth rate of the Commonwealth in the 1950s was 3.2 percent, while the national average was 18.5 percent. In the 1960s Kentucky's growth rate was 5.9 percent as compared to the national rate of 13.3 percent. Recent Census Bureau estimates indicate that Kentucky had an even more rapid population growth rate from 1970 to 1975. Kentucky's greater growth in recent years is associated with increasing urbanization and the revival of the coal industry. Both may portend future trends.

Urbanization has been one of the most important long-term trends in the distribution of Kentucky's population **(Table 2)**. Using the Census Bureau's definition of urban, which includes all residents of incorporated and unincorporated places of over 2,500, Kentucky's population was 52.3 percent urban in 1970, for the first time predominantly urban. The state's first urban population appeared on the census of 1810, and the proportion of the population that was urban has increased in every decade since then except

for the Depression decade of the 1930s. (For more detail on urban growth see **Maps 48–54**.)

It appears that in the future a greater share of Kentucky's people will live in counties that the Census Bureau designates as parts of Standard Metropolitan Statistical Areas (SMSAs). Each SMSA contains a central city or cities of over 50,000 and the surrounding counties which meet Census Bureau criteria of metropolitanization, based on population density, commuting patterns, and other factors. In 1970 eight of Kentucky's 120 counties were included in SMSAs. Between 1970 and 1974 nine more counties were added to SMSAs, so that by 1974 seventeen Kentucky counties, containing 1,510,595 people, were considered metropolitan **(Table 3)**. In future more counties will surely be added to SMSAs and these areas will contain an increasing share of the state's population. A more "metropolitanized" outlook is likely to become evident in all aspects of state life, from economics to politics.

The distribution of population in the state **(Map 4)** reflects the dichotomy between urban and other incorporated places, shown by circles, and unincorporated areas, shown by shading. The rural population is densest in the Bluegrass Region, in the eastern and western coal fields, and around the major cities, notably Louisville, Lexington, Covington-Newport, Ashland, Paducah, and Owensboro. Many of the rural persons in the coal fields and near urban areas are what is known as "rural nonfarm" population. That is, while they live in rural areas they are not in agricultural occupations. Around major urban centers, these rural nonfarm persons

Most of the population of Kentucky is now concentrated in urban areas. Economic growth and cultural attractions, such as the ethnic festival in Louisville shown at left, have been major forces in attracting population from rural areas. Even small cities like McKee (*far left*) registered modest growth in the 1960s, but Jackson County, of which McKee is the seat, lost 6.3 percent of its population.

TABLE 1
Population Growth, 1790-1970

Year	Population	Percentage of growth from previous census	
		Kentucky	U.S.
1790	73,677	—	—
1800	220,955	199.9	35.1
1810	406,511	84.0	36.4
1820	564,317	38.8	33.1
1830	687,917	21.9	33.5
1840	779,828	13.4	32.7
1850	982,405	26.0	35.9
1860	1,155,684	17.6	35.6
1870	1,321,011	14.3	22.6
1880	1,648,690	24.8	30.2
1890	1,858,635	12.7	25.5
1900	2,147,174	15.5	21.0
1910	2,289,905	6.6	21.0
1920	2,416,630	5.5	15.0
1930	2,614,589	8.2	16.2
1940	2,845,627	8.8	7.3
1950	2,944,806	3.5	14.5
1960	3,038,156	3.2	18.5
1970	3,218,706	5.9	13.3

TABLE 2
Urban and Rural Populations, 1790-1970

Year	Urban population			Rural population	
	Number	Percentage of change from previous census	Percentage of total	Number	Percentage of change from previous census
1790	—	—	—	73,677	—
1800	—	—	—	220,955	+199.9
1810	4,326	—	1.1	402,185	+ 82.0
1820	9,291	+114.8	1.6	555,026	+ 38.0
1830	16,367	+ 76.2	2.4	671,550	+ 21.0
1840	30,948	+ 89.1	4.0	748,880	+ 11.5
1850	73,804	+138.5	7.5	908,601	+ 21.3
1860	120,624	+ 63.4	10.4	1,035,060	+ 13.9
1870	195,896	+ 62.4	14.8	1,125,115	+ 8.7
1880	249,923	+ 27.6	15.2	1,398,767	+ 24.3
1890	356,713	+ 42.7	19.2	1,501,922	+ 7.4
1900	467,668	+ 31.1	21.8	1,679,506	+ 11.8
1910	555,442	+ 18.8	24.3	1,734,403	+ 3.3
1920	633,543	+ 14.1	26.2	1,783,087	+ 2.8
1930	799,026	+ 26.1	30.6	1,815,563	+ 1.8
1940	849,327	+ 6.3	29.8	1,996,300	+ 10.0
1950	1,084,070	+ 27.6	36.8	1,860,736	− 6.8
1960	1,353,215	+ 24.8	44.5	1,684,941	− 9.4
1970	1,684,053	+ 24.4	52.3	1,534,653	− 8.9

Covington (*right*), settled in 1791, is part of the northern Kentucky urban area, the second most populous in the state. Cincinnati, directly across the Ohio River, is the major urban focus of the area.

Most of Kentucky's small communities are served by their own post offices, such as this one at Melber in the Jackson Purchase (*far right*). It is typical of small hamlets in many rural areas of the state.

TABLE 3
Kentucky Standard Metropolitan Statistical Areas
(*As defined by the United States Office of Management and Budget, July 1, 1974*)

Area and Kentucky counties included	1970 population
Cincinnati, Ohio-Ky. (Boone, Campbell, Kenton)	250,753
Clarksville-Hopkinsville, Ky.-Tenn. (Christian)	56,224
Evansville, Ind.-Ky. (Henderson)	36,031
Huntington-Ashland, W.Va.-Ky.-Ohio (Boyd, Greenup)	85,568
Lexington-Fayette (Bourbon, Clark, Fayette, Jessamine, Scott, Woodford)	266,701
Louisville (Bullitt, Jefferson, Oldham)	735,832
Owensboro (Daviess)	79,486
Total population in SMSAs	1,510,595
Percentage of state population	46.9

POPULATION DISTRIBUTION

MAP 4

Population of incorporated places

o	•	◉	●	
54–499	500–999	1000–2499	2500–9999	Places of 10,000 or more are shown by graduated circles

——— Urbanized areas are outlined

Persons per square mile

| 52 and more | 40–51 | 28–39 | 20–27 | 19 and less |

25 0 25 50 MILES
25 0 25 50 KILOMETERS

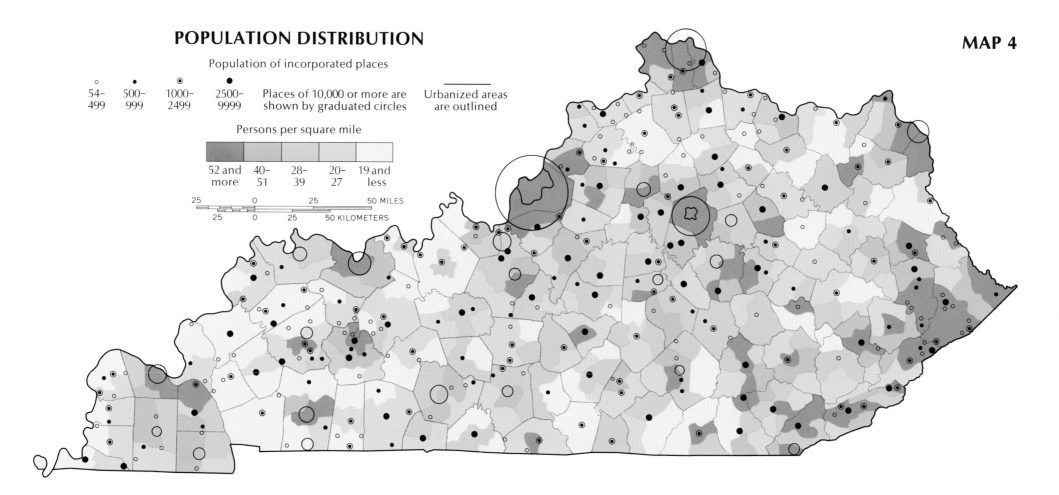

are commuters **(Map 55)** who work in cities ten, twenty, or even fifty miles from their homes. In the coal field areas, the rural non-farm population works in mining and service occupations at scattered places in the areas. Much of the coal fields' "rural" population lives in unincorporated company towns and string towns, especially in the eastern part of the state. The lowest densities of rural population are found in several areas of the state: the western portion of the Pennyroyal between Bowling Green and Paducah; the Mountains in areas with little mining; portions of the Western Coal Field where coal is not actively being mined; and along the floodplain of the Mississippi River at the western end of the state.

The six largest metropolitan centers are defined by the Census Bureau as Standard Metropolitan Statistical Areas. Smaller urban centers with populations of 2,500 to 35,000 are scattered unevenly across the state. Most of these centers were originally agricultural

marketing and retail centers for the surrounding rural areas. These towns are most numerous in the Bluegrass and least numerous in the Mountains **(Map 54)**. In addition to marketing and retailing, many of these towns have industrial or other functions. Bowling Green, Morehead, Murray, and Richmond have state universities, while Frankfort is the state capital. The coal fields have many mining towns, such as Madisonville, Earlington, Middlesboro, and Harlan, in addition to scattered rural population. Along the Ohio, Tennessee, and Mississippi rivers are port towns, including Paducah, Maysville, Carrollton, Calvert City, Hawesville, and Lewisport, and towns which have recently grown as a result of the waterways and proximity to coal and power. Towns near Kentucky's large recreational lakes have grown as service centers for the tourists and seasonal populations of these areas, especially around Kentucky, Barkley, and Cumberland lakes.

MAP 5

POPULATION CHANGE 1950 TO 1960

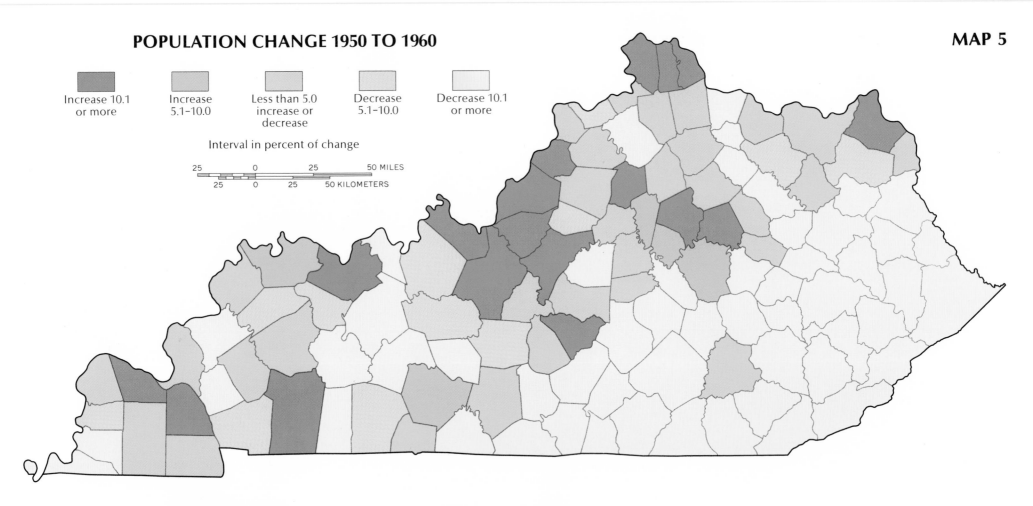

| Increase 10.1 or more | Increase 5.1–10.0 | Less than 5.0 increase or decrease | Decrease 5.1–10.0 | Decrease 10.1 or more |

Interval in percent of change

25   0   25   50 MILES
25   0   25   50 KILOMETERS

Harlan, an important coal-mining center in southeastern Kentucky. The population of Harlan County declined drastically in the 1960s when nationwide use of coal forced many miners to seek employment in northern industrial cities. The county's population has increased in recent years, however.

The general changes in population distribution in the Commonwealth from 1950 to 1960 **(Map 5)** and from 1960 to 1970 **(Map 6)** indicate a continuation of long-term trends toward the growth of metropolitan centers. Population growth has been especially rapid since 1950 in the Lexington, Louisville, and Covington-Newport areas. Bullitt County, near Louisville, registered a 65.9 percent population gain from 1960 to 1970, while Fayette County (Lexington) grew 32.2 percent and Boone County (near Covington-Newport) increased 49.6 percent in the same period. The largest population losses in both decades were in the eastern coal mining counties. Harlan County lost 26.9 percent of its population from 1960 to 1970, although neighboring Leslie County increased. Recent Census Bureau estimates show, however, that the coal mining counties have registered dramatic population increases since 1970, reflecting increased use of coal and a boom in the coal mining industry.

POPULATION CHANGE 1960 TO 1970

MAP 6

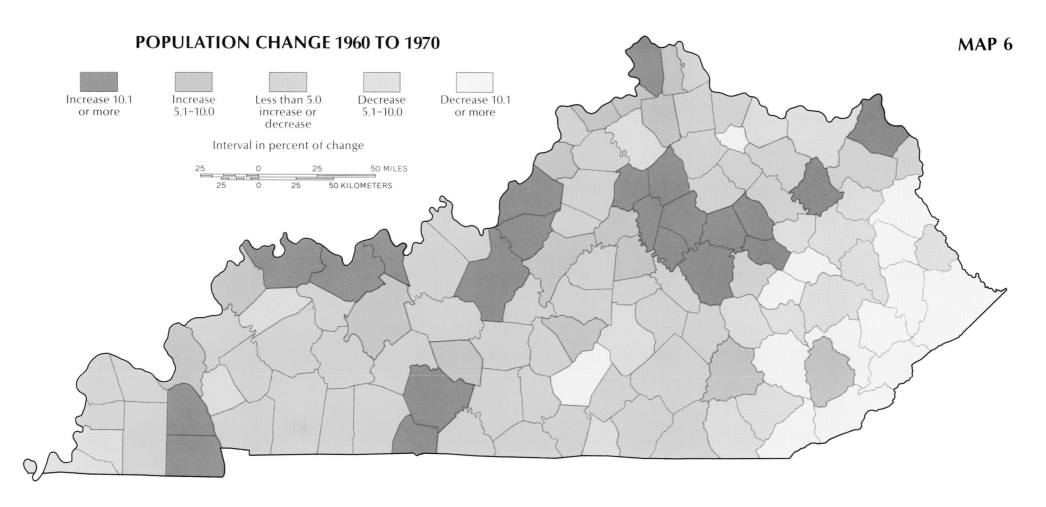

Increase 10.1 or more

Increase 5.1-10.0

Less than 5.0 increase or decrease

Decrease 5.1-10.0

Decrease 10.1 or more

Interval in percent of change

25   0   25   50 MILES
25   0   25   50 KILOMETERS

Shifts in population from rural to urban are exemplified in these two aerial views of the south side of Lexington, taken in 1959 (*far left*) and 1975. Note the new suburban housing developments and the growth of large shopping centers near the intersection of Nicholasville and New Circle roads, examples of poorly planned urban development.

Distribution, Density,
& Growth   21

MAP 7

POPULATION AGE STRUCTURE, AGES 18-64

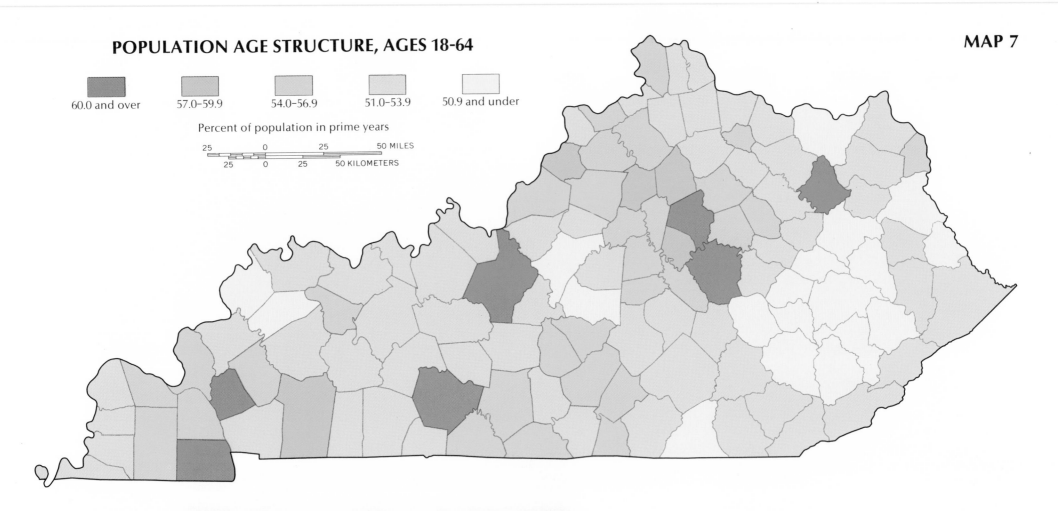

60.0 and over   57.0–59.9   54.0–56.9   51.0–53.9   50.9 and under

Percent of population in prime years

25   0   25   50 MILES

25   0   25   50 KILOMETERS

Kentucky has a higher proportion of persons under eighteen years of age than the national average. The percentage of older persons is also greater. Many persons in the "prime years" left the state in the 1960s, seeking better employment opportunities elsewhere. This trend appears to have slowed since 1970.

Compared to the United States as a whole, Kentucky has a slightly smaller proportion of population in what might be called the "prime years," ages eighteen through sixty-four (**Map 7**). Conversely, the state has proportionately more people under eighteen and over sixty-four, probably because many Kentuckians in the prime years have moved away. Areas with the greatest proportion of the population in the prime years are mainly those around urban and educational centers. Regionally, most of the counties in the Mountains have less than 51 percent of the population in the prime years. The socioeconomic effect of this young-and-old disproportion is to place a greater burden on the state's educational and welfare institutions.

PHILLIP D. PHILLIPS and THOMAS P. FIELD

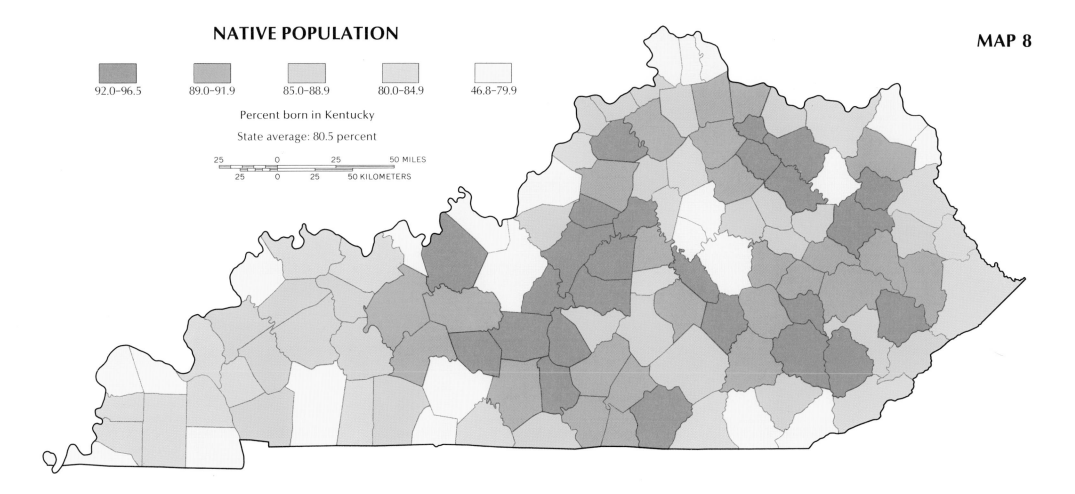

NATIVE POPULATION

92.0-96.5   89.0-91.9   85.0-88.9   80.0-84.9   46.8-79.9

Percent born in Kentucky

State average: 80.5 percent

25   0   25   50 MILES

25   0   25   50 KILOMETERS

MAP 8

MIGRATION AND RESIDENTIAL PREFERENCE PATTERNS

MAPS 8-12

Four out of five Kentuckians are Kentuckians by birth, but regional variations within the Commonwealth are striking **(Map 8).** At the county level, Kentucky-born population varies from as low as 46.8 percent to as high as 96.5 percent. In twenty-three counties the range is from 46.8 to 79.9 percent, all below the state's 80.5 percent average. These counties include most of the large urban centers, the Fort Knox and Fort Campbell military bases, and seven of the eight state-supported colleges and universities. Several border counties lack the foregoing associations but have above-average proportions of people born outside Kentucky. Counties near the state border with urban associations and relatively high proportions of persons from outside of Kentucky include Jefferson County

(Louisville), McCracken County (Paducah), Bell County (Middlesboro), Boyd and Greenup counties (Ashland), and the three northernmost counties (Boone, Kenton with Covington, and Campbell) within Cincinnati's metropolitan area. Interior counties and their urban areas which have low proportions of Kentucky-born people are Fayette and Jessamine counties (Lexington), Madison County (Richmond), and Warren County (Bowling Green).

Some of these same areas include major state universities with sizable numbers of people born outside Kentucky. Among these are Rowan County (Morehead State University); Madison County (Eastern Kentucky University and the private Berea College); Warren County (Western Kentucky University); and Calloway County

The influx of population into urban areas has been partially absorbed by the development of trailer parks, such as this one in Franklin County.

(Murray State University). Fort Knox in Hardin and Meade counties and Fort Campbell in Christian County (which also includes the city of Hopkinsville), as military bases, have high proportions of population born outside Kentucky.

The native-born population is highest in the predominantly rural counties of central and eastern Kentucky encircling the Inner Bluegrass. The two largest blocks are formed by a southwest-northeast cluster of ten counties east and south of Louisville to Bowling Green, and a six-county cluster in the northeast around Rowan County (Morehead). If all counties having 85 percent or more of their populations Kentucky-born are included, the pattern can be viewed as covering nearly all areas of Kentucky.

Kentuckians, like Americans in general, are mobile. One of the ways of measuring this mobility is through changes in household location. Between 1965 and 1970, more than 46 percent of the Kentucky population five years of age or older changed household locations. About one-third of this group, 14.5 percent of the population five years or older, migrated into a new county of residence (**Map 9**). This was a slightly higher level of mobility than was found for the 1955–1960 period, when 14.1 percent moved to a new county.

The 1970 pattern of migrant population by counties is a highly diverse one. At one extreme, Wayne County in the southeastern Pennyroyal had less than 5 percent of its population in-migrant. Eight other counties, all in the southeastern Pennyroyal or in the Mountains, had in-migrant populations of under 7 percent. In striking contrast, Hardin County, south of Louisville and including several growing urban centers plus the Fort Knox military base, had an in-migrant figure of 42.3 percent. Eight other counties spread across the state had percentages greater than 25.

The twenty-five counties having the highest level are widely scattered across the state, though only the westernmost fringes of the Mountains are included. The counties in this highest range have one or more of several attracting factors, such as employment opportunities, shopping and recreational facilities, and absence of environmental problems. The largest block of counties is in the central Bluegrass area of Lexington and its satellite cities, plus the state capital of Frankfort in Franklin County, which also has Kentucky State University. A second, more widespread group of counties has institutions of higher learning which expanded dramatically in the late 1960s: Rowan, Madison, Warren, and Calloway counties. In the military installations of Fort Knox in Hardin and Meade counties and Fort Campbell in Christian County frequent turnovers resulted in high in- and out-migration proportions.

MIGRANT POPULATION

MAP 9

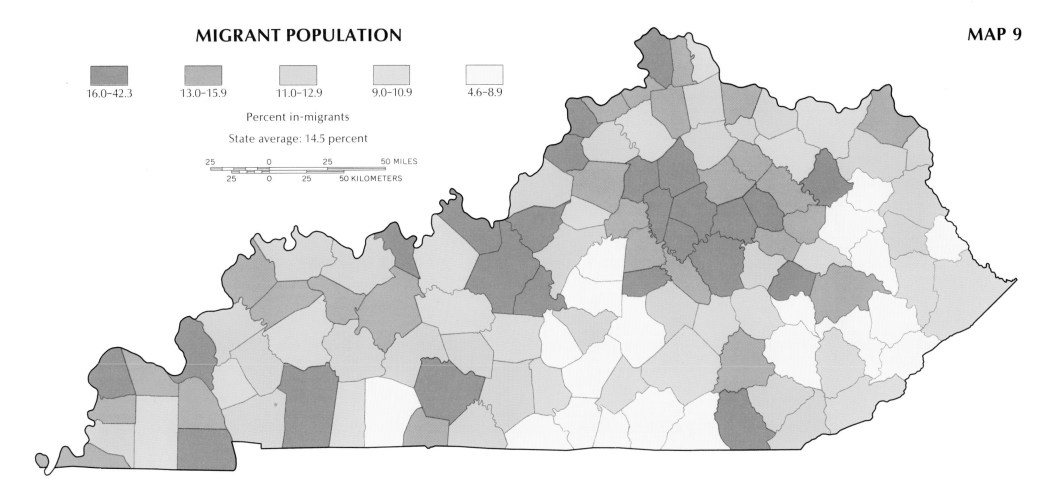

16.0–42.3   13.0–15.9   11.0–12.9   9.0–10.9   4.6–8.9

Percent in-migrants

State average: 14.5 percent

25   0   25   50 MILES

25   0   25   50 KILOMETERS

A fourth high in-migration association, typical for a considerable number of counties in Kentucky, is location on the fringes of large, expanding metropolitan areas served by increasingly good transportation arteries: the areas around Fayette County, the counties north and south of Louisville and Jefferson County, and Boone County in northern Kentucky within the Cincinnati–Covington metropolitan area. Some predominantly rural counties have also had recent surges of in-migration to provide labor for newly established industries. The two principal examples are Hancock County on the Ohio River, with a large new aluminum reduction plant, and Livingston County, which has new chemical industries associated with the completion of Barkley Dam and its hydroelectric production on the lower Cumberland River. Both counties are served by excellent river transportation and the nearby Western Coal Field.

Other factors attracting in-migrants include increased transportational accessibility, as in the areas along Interstate Route 71 in Carroll and Gallatin counties in northern Kentucky; developing resources, as in the Western Coal Field or the recreation-oriented counties in the Jackson Purchase; and spreading urban growth, either because of expanding urban centers, as in Somerset, or because of metropolitan fringe growth into formerly more rural counties, as in Greenup County northwest of Ashland or Shelby County near Louisville. In-migrants have also contributed to the suburbanization of McCracken County outside of Paducah.

Two sizable blocks of counties have the lowest proportions of in-migrants, one in the southeastern Pennyroyal and one in the Mountains. Several other counties have almost as low levels. The general lack of expanding urban or industrial opportunities has been typical of these counties.

Migration & Residential
Preference 25

Bowling Green (*above*) is the largest urban center in the Pennyroyal. Its diversified industries and the presence of Western Kentucky University have made it attractive to in-migration. A statewide survey found Bowling Green, Owensboro, and the Bluegrass to be the most preferred residential areas in the state. Other counties with large state universities, such as Eastern Kentucky University at Richmond (*below*), have experienced similar growth.

The smallest areal units in Kentucky between which migration flows are measured are State Economic Areas (SEAs). Almost 156,000 persons migrated across SEA boundaries at least once between 1965 and 1970. State Economic Areas are defined by the United States Census Bureau as relatively homogeneous subdivisions of states consisting of single counties or groups of counties which have similar economic and social characteristics. The boundaries of these fourteen State Economic Areas in Kentucky have been drawn by the Census Bureau in such a way that each area has certain characteristics which distinguish it from adjoining areas.

Sixteen migration flows between SEAs exceeded 2,500 persons in 1965–1970 **(Map 10)**. Although the Appalachian portion of Kentucky is well known as a source region for migrations to manufacturing centers outside the state, the greatest migration volumes within Kentucky occurred in the north and north-central areas of the state. The far western and south-central regions experienced no absolute volumes greater than 2,500 persons.

Several pairs of regions traded migrants. This happened primarily around the large urban areas. Louisville (SEA #10) traded with SEA #3 to the south and SEA #6 to the east. Covington (SEA #11) also traded with SEA #6, while Lexington (SEA #14) exchanged with surrounding SEA #7. A step-down type of migration occurred from the eastern to the central portion of the state. SEA #9 gave migrants to both SEA #8 and SEA #6, while SEA #8 donated migrants to SEA #6. The probability of large absolute flows of migrants is greatest where population is densest. Because the northern and central portions of Kentucky are highly populated, the flows depicted in **Map 10** were not unusual.

Generally large cities attract more in-migrants because of increased job opportunities, urban services, and diversified amenities. Some Kentucky cities seem to be preferred over others, however, for they attract more in-migrants. A map of in-migration attraction values per 1,000 inhabitants for twenty-five cities (**Map 11**), computed from Census Bureau tabulations of 1970 residents who lived elsewhere in 1965, shows that the central part of the state is the strongest attraction area. Frankfort, Danville, Lexington, and Winchester, cities with strong attraction values, are all within the Bluegrass area. Elizabethtown and Bowling Green also exhibit strong attractions. Louisville shows very little attraction strength per 1,000 inhabitants. Cities of poorest attraction lie at the areal extremities of the state. Covington and Ashland to the north and northeast are examples, as are Mayfield in the extreme western portion of the state, Monticello at the extreme south, and Hazard to the southeast.

A comparison of the pattern of urban in-migration attraction (**Map 11**) with the residential preference pattern (**Map 12**) reveals some striking similarities. The in-migration attraction of an urban area is strongly related to an aggregate residential preference value attributed to that city by potential migrants. Over 1,500 individuals from twenty-five Kentucky cities were surveyed by mail and asked to evaluate the residential preference of all other survey cities. Respondents indicated whether they thought each city was a very desirable, desirable, fair, undesirable, or very undesirable place in which to live. They could also indicate that they lacked sufficient knowledge to evaluate the city. Responses from the 435 individuals who replied to the survey were used in preparing **Map 12.** The age and income characteristics of respondents were generally representative of statewide characteristics, although the sample is biased towards persons with stronger educational backgrounds.

The Inner Bluegrass emerged as the most preferred residential area in the Commonwealth. From the Inner Bluegrass, preference decreased rapidly to the north, south, and east. The west and west-central areas maintained a high degree of residential preference, especially around Owensboro and Bowling Green. This pattern is generally in keeping with that of in-migration attraction as shown on **Map 11.**

STEPHEN E. WHITE, WILLIAM A. WITHINGTON, and RICHARD I. TOWBER

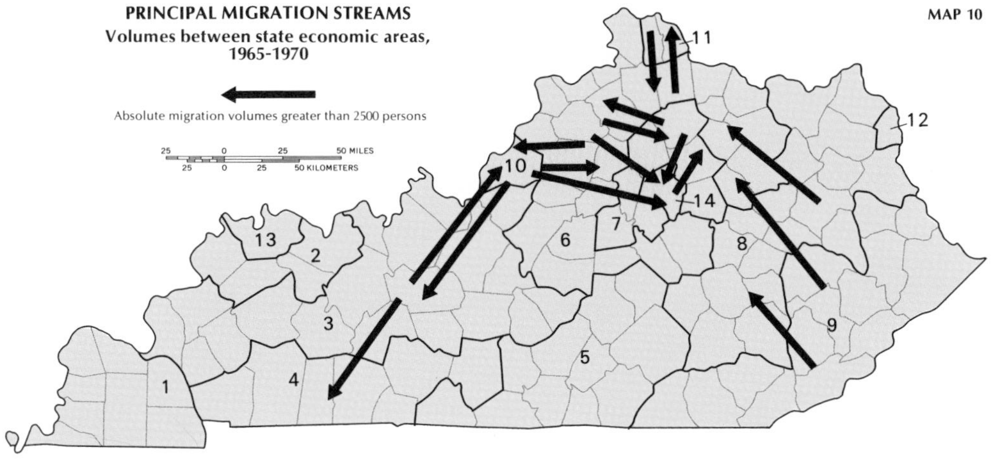

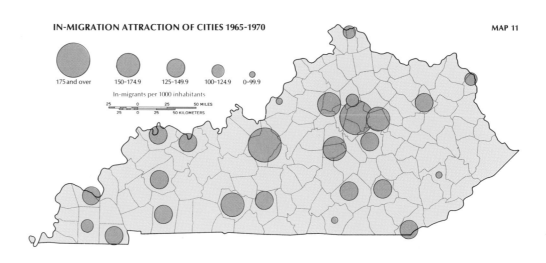

PRINCIPAL MIGRATION STREAMS
Volumes between state economic areas,
1965-1970

MAP 10

Absolute migration volumes greater than 2500 persons

IN-MIGRATION ATTRACTION OF CITIES 1965-1970

MAP 11

175 and over   150-174.9   125-149.9   100-124.9   0-99.9

In-migrants per 1000 inhabitants

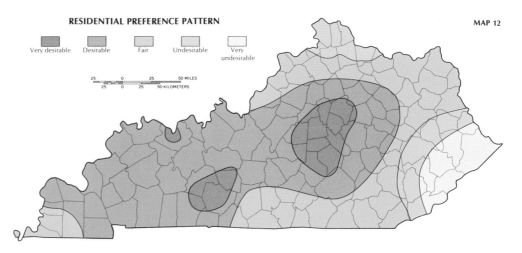

RESIDENTIAL PREFERENCE PATTERN

MAP 12

Very desirable   Desirable   Fair   Undesirable   Very undesirable

27

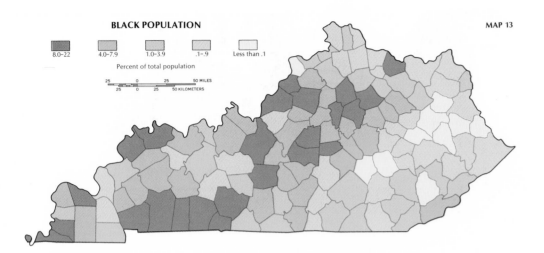

8.0–22   4.0–7.9   1.0–3.9   .1–.9   Less than .1

Percent of total population

25   0   25   50 MILES
25   0   25   50 KILOMETERS

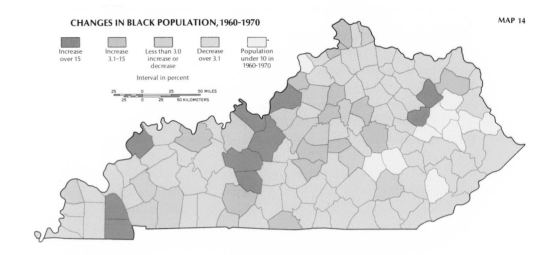

Increase over 15   Increase 3.1–15   Less than 3.0 increase or decrease   Decrease over 3.1   Population under 10 in 1960–1970

Interval in percent

25   0   25   50 MILES
25   0   25   50 KILOMETERS

MAPS 13 and 14   BLACK POPULATION

Early black settlers to Kentucky came from the East, primarily Virginia, Maryland, and the Carolinas. Black slaves were brought by planters and by Irish, Scotch, and German immigrants at the time of the first permanent settlements at Harrodsburg and Boonesborough in 1775. The planters migrated to the Bluegrass initially and later to the western portions of the state, bringing their slaves with them. By 1790 the black population of Kentucky was almost 12,000, over 15 percent of the total population. Many plantation owners left Kentucky after 1830, taking their slaves south, where the growing of cotton was more profitable than the production of hemp, cereals, or livestock. The institution of slavery in Kentucky was opposed by activists such as Robert Breckinridge, John Fee, Cassius Clay, and J. R. Underwood, yet slave ownership continued until the Civil War.

After the Civil War, blacks left the farms and began moving to the cities. In the two decades after the war, the black population increased slowly and reached a peak of 284,706 in 1900. After 1900, however, each decennial census tabulated fewer blacks until 1960, when the black population again began to increase slowly.

In 1970, about 7 percent of Kentucky's population was black. Kentucky is often viewed as a southern state. Its black population, however, is substantially lower than that of Tennessee (15.8 percent), Alabama (26.2 percent), Georgia (35.0 percent), Mississippi (36.8 percent), or Louisiana (42.7 percent). In its proportion of black population Kentucky is more akin to neighboring states to the north: Indiana (6.9 percent) and Ohio (9.1 percent). Kentucky's blacks, like those in the northern states, live primarily in urban areas (Map 13) instead of in the rural countryside as do most southern blacks. In 1970, about 80 percent of Kentucky's blacks lived in urban areas of 2,500 people or more, whereas 50 percent of the whites lived in towns and cities. Only about 4 percent of the black population lived on farms, mainly in some Purchase and Pennyroyal counties. About 50 percent of all Kentucky blacks lived in the state's two largest cities, Louisville and Lexington, and, as elsewhere, most metropolitan blacks lived in inner cities (Map 61).

A hundred years ago the Bluegrass, especially the fertile Bluegrass counties surrounding Lexington, had one of the highest proportions of black population in the state. This had declined to about 8 percent by 1970, although the black population in Lexington was about 12 percent. Few of the Mountain counties have ever had many blacks. In 1970 blacks were less than 2 percent of the Mountain population. In the rural central and western regions of the state, the black population has traditionally been low. Most counties in the Pennyroyal, the Western Coal Field, and the Jackson Purchase had fewer than 1,000 blacks each in 1970. One exception was Christian County in the western Pennyroyal, where 22

percent of the population was black. In Hopkinsville, the county seat, 21.9 percent were black. These are the highest proportions of black people in the state. The reason for the high concentration seems to be twofold. First, part of Fort Campbell Army Base lies in the extreme southern part of Christian County. More importantly, Hopkinsville sits astride one of the principal migration streams of blacks moving from the Deep South to northern cities. Blacks from Alabama move northward to Nashville, then to Hopkinsville, Evansville, Indianapolis, and on perhaps to Chicago or Detroit. Other concentrations of blacks in Kentucky are associated with state universities (Murray State University in Calloway County and Western Kentucky University in Warren County), military bases (Fort Knox in Hardin County), or Job Corps centers in Union and Edmonson counties.

Between 1960 and 1970 Kentucky's black population generally continued to increase in counties with industrial towns and service centers **(Map 14)**. The growth in black population was the result of both natural increase and in-migration to towns with employment opportunities in "blue collar" jobs, which accounted for about two-thirds of all employed black workers in 1970. Most rural counties recorded a decrease in black population.

KARL B. RAITZ and WILFORD A. BLADEN

Most of Kentucky's black people have migrated from rural to urban areas seeking better economic opportunities and urban amenities, such as the recreational facilities of city parks (Louisville, *left*) and large churches (Lexington, *above left*). Kentucky's black children have better education opportunities today than in the past. *Above:* A group of high school student editors attend a journalism workshop at the University of Kentucky.

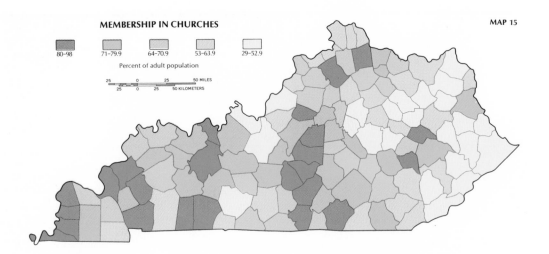

80-98 71-79.9 64-70.9 53-63.9 29-52.9

Percent of adult population

25   0   25   50 MILES
25   0   25   50 KILOMETERS

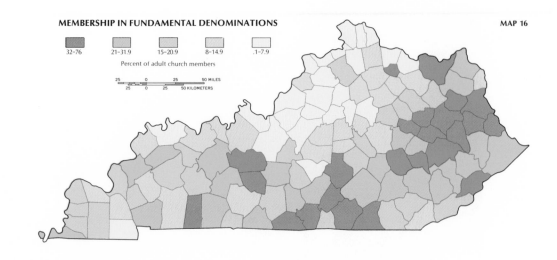

32-76 21-31.9 15-20.9 8-14.9 .1-7.9

Percent of adult church members

25   0   25   50 MILES
25   0   25   50 KILOMETERS

MAPS 15 and 16

CHURCH AFFILIATIONS

In 1971 Kentucky had almost 9,000 churches. Membership in these churches totaled 1,412,000, or about 60 percent of the adult population (those fourteen years and older). Counties vary greatly in the percentage of the adult population reported as church members **(Map 15)**. Church membership is highest in those counties where settlement is stable and where strongly organized denominations are well represented, particularly in rural areas of the Pennyroyal, Western Coal Field, and Purchase. It is lowest in those counties with heavy in-migration and thus a temporary loosening of church associations or in those which are relatively inaccessible, thinly settled, or characterized by low incomes. The low ranking of counties such as Jefferson, Hardin, Warren, Fayette, Jessamine, and Madison is readily explained by heavy in-migration. The low percentages reported for the relatively poor, isolated, and thinly occupied areas of rugged terrain in the Mountains probably reflect the difficulties for actual or potential congregations, but they also quite probably indicate the relative importance in the Mountains of autonomous, rather loosely organized church groups, largely oriented toward the fundamentalist denominations.

Early in-migrations to the state influenced denominational distribution. For example, the Catholic population, which represents 16.1 percent of the state's church members, is concentrated in counties having early German settlements, such as those along the Ohio River and several of the central counties, especially Marion, Nelson, Hart, and Washington.

The Southern Baptists, the largest single denomination, with 34.1 percent of the state's church membership, are concentrated in the rural areas, except in the Mountains, where more conservative denominations prevail. Methodism (14.3 percent of the church membership) is dispersed throughout the state except for the more remote Mountain counties. Christian Churches and Disciples of Christ (4.2 percent) and Churches of Christ (4.0 percent) are strongest in northeast-central Kentucky. Presbyterians (1.7 percent) are most numerous in the central and western parts of the state.

Twenty-one percent of Kentucky's church members belong to fundamental (conservative) denominations **(Map 16)**. Included as fundamental are groups such as the Church of God (1.3 percent); Nazarenes (.8 percent); Independent Churches of Christ; Primitive, United, and Old Regular Baptists; Jehovah's Witnesses; Assemblies of God; Pentecostal and Holiness groups; and other smaller sects. Higher percentages of membership in fundamental churches correspond to the poorest eastern counties and those of south-central Kentucky, while those having the least fundamental membership are the urban and more prosperous north-central counties.

KARL B. RAITZ and RICHARD BOOTH

Kentucky had almost 9,000 churches in 1971.
Southern Baptist churches, like that near Mayfield
in the Purchase (*above left*), represent the single
largest denomination in the state. Lystra Church of
Christ (*far left*) is an example of a smaller
denomination concentrated in northeast-central
Kentucky. The Apostolic Faith Church of God in
Haymond (*left*) is one of many fundamentalist
congregations found mainly in the Mountains.
Kentucky's Catholic churches, such as the Cathedral
Basilica of the Assumption in Covington (*above*),
are most numerous in urban areas along the Ohio.

IV. SOCIAL AND ECONOMIC PATTERNS

MAPS 17-19

SOCIOECONOMIC CHARACTERISTICS

Twenty-four variables relating to income, education, housing, and occupation for Kentucky counties were employed in a statistical technique called "factor analysis" to produce **Map 17.** The standard deviation portrayed for each county is the measure of how much its value differs from the mean socioeconomic scale for the entire state. The analysis indicates that the highest socioeconomic levels are found in the most metropolitan counties, including Fayette (Lexington); Jefferson, Bullitt, and Oldham (Louisville); Boone, Kenton, and Campbell (Covington-Newport); Boyd and Greenup (Ashland); and Daviess (Owensboro). High levels are also found in other portions of the Bluegrass and in the Jackson Purchase. The lowest levels are in southeastern and portions of eastern Kentucky. The impact of Lake Cumberland and Interstate Route 75 can be seen in the somewhat higher levels of Pulaski, Rockcastle, Laurel, and Whitley counties than of surrounding counties. The general income and education levels of Kentucky as a whole are lower than the national average. While Kentucky's highest levels are above the national norm, the lowest levels are among the lowest in the nation.

According to official government definitions, many Kentuckians fall below the minimum standards of income and are classified as poor. The poverty threshold varies with the size, composition, and residence (urban or rural) of a family, ranging from a low of $1,487 for a rural female over sixty-five years of age to $6,116 for a nonfarm family of seven or more with a male head. The poverty threshold for a nonfarm family of four headed by a male was $3,745. The poverty threshold definitions used here are based on 1969

Older houses on Fifth Street in Covington. The condition of housing in various parts of an urban area reflects the levels of social and economic development.

MAP 17

SOCIOECONOMIC LEVELS

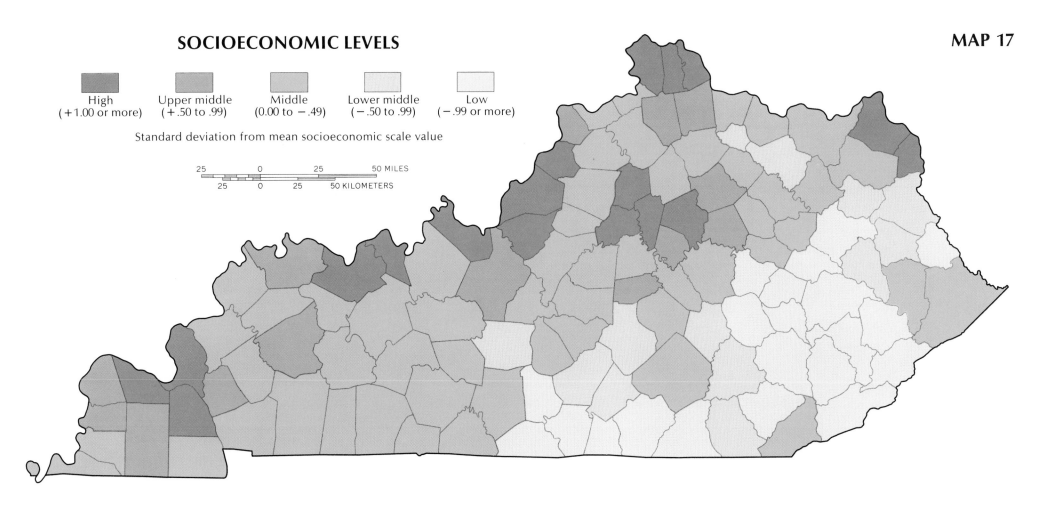

High
(+1.00 or more)

Upper middle
(+.50 to .99)

Middle
(0.00 to −.49)

Lower middle
(−.50 to .99)

Low
(−.99 or more)

Standard deviation from mean socioeconomic scale value

25   0   25   50 MILES

25   0   25   50 KILOMETERS

Socioeconomic contrast is evident in these two views of Chevrolet, a coal-mining "company town" in Appalachian Kentucky. *Far left:* Coal miners' dwellings; similar housing can be found clustered along many branches, creeks, and hollows in eastern Kentucky. *Left:* The home of a coal company executive.

MAP 18

POPULATION BELOW POVERTY LEVEL

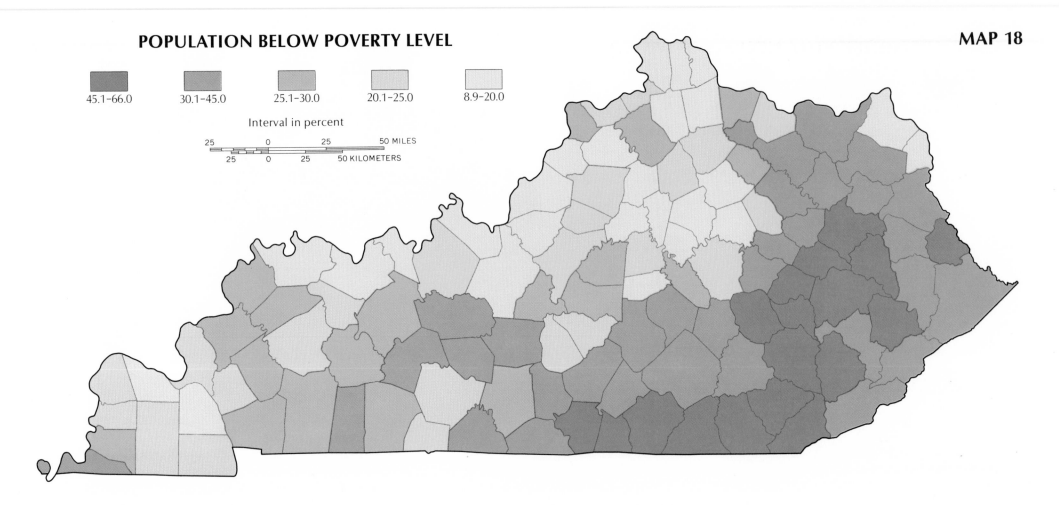

45.1–66.0   30.1–45.0   25.1–30.0   20.1–25.0   8.9–20.0

Interval in percent

25   0   25   50 MILES
25   0   25   50 KILOMETERS

income adjusted to take inflation into account. The distribution of Kentuckians below the poverty level varies widely by county (**Map 18**), from a low of 8.9 percent to a high of 66.0 percent. The highest percentage levels of poverty are found in rural areas, although larger numbers may exist in metropolitan centers. The poorest areas of Kentucky are in the eastern Mountains and the hillier portions of the Pennyroyal. The smallest proportions of persons in poverty are found in the metropolitan areas and in nonmetropolitan portions of the Bluegrass, in the Jackson Purchase, and in the counties bordering the Ohio River from Covington-Newport to Henderson.

Income statistics for 1974 released by the U.S. Bureau of Economic Analysis in June 1976 indicate that all counties have shared in the state's gain in per capita income, but the geographic distribution of poverty counties has not changed since 1970. For in-

stance, McCreary County's per capita income in 1974 was $2,045, the lowest of the 120 counties and far behind Jefferson County's top-of-the-list $5,653. Although McCreary County's per capita income had increased 71 percent from the 1970 level, it was still the state's poorest county. Some other eastern Kentucky counties nearly as poor in 1970 have fared better under the coal boom of recent years. For example, Martin County, once reputed to be one of the poorest counties in the United States, had a per capita income jump from $1,492 in 1970 to $3,571 in 1974, a gain of 139.3 percent. Other coal counties have done nearly as well. Knott County jumped from a per capita level of $1,365 in 1970 to $2,693 in 1974. Leslie County nearly doubled its income during the period, from $1,288 to $2,298. Perry County jumped from $2,327 to $4,000. Six counties climbed above the $5,000 per capita income level in 1974 for the first time and now nine are above this level, headed by

MAP 19

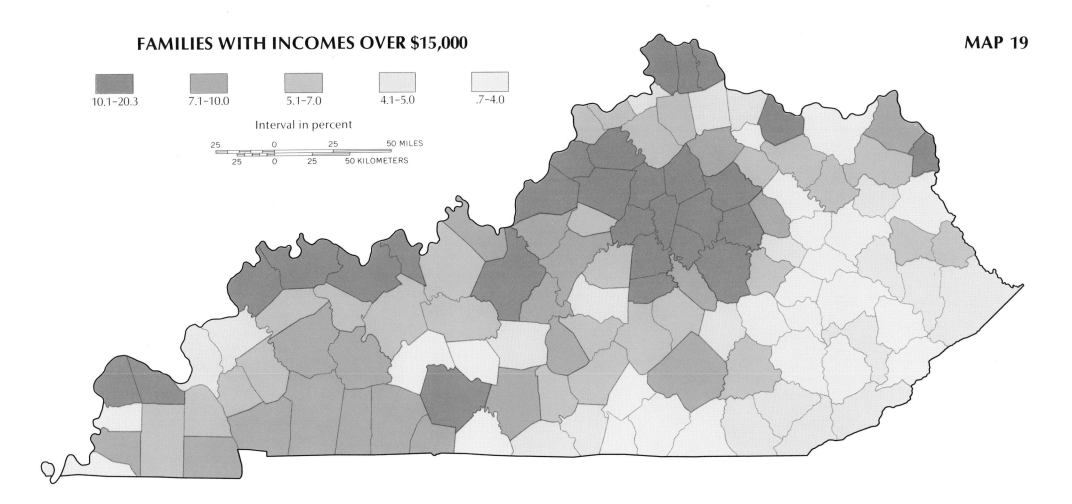

10.1–20.3	7.1–10.0	5.1–7.0	4.1–5.0	.7–4.0

Interval in percent

25 0 25 50 MILES
25 0 25 50 KILOMETERS

Jefferson County with $5,653. The other counties are: Woodford ($5,507); Clark ($5,348); Fayette ($5,287); Hopkins ($5,248); Campbell ($5,156); Kenton ($5,118); and Franklin ($5,067).

Families with incomes over $15,000 in 1969 (**Map 19**) are concentrated in metropolitan areas, in terms of both absolute numbers and percentages of the population. The Bluegrass in particular stands out as a high-income region. The lowest proportions of families with incomes over $15,000 are found in the eastern Mountains. An interesting exception to the general pattern in the Mountains is Pulaski County, which has a larger proportion of high-income persons than any of the surrounding counties. The Pulaski County seat, Somerset, is a regional trade center and the focus of recreational and residential development around Lake Cumberland.

PHILLIP D. PHILLIPS and WILFORD A. BLADEN

The cost of homes directly affects the composition of population in an area. Suburban houses in large Kentucky cities, like this one in Lexington, are beyond the reach of all but high and upper middle income groups.

The modern facilities of the Kentucky State Police headquarters in Frankfort (*above*) serve the entire Commonwealth and assist local law enforcement agencies. *Right:* A state trooper helps a motorist.

In recent years areas of denser urban development, especially those with a high proportion of rental units, have recorded higher crime rates. Two policemen on patrol in downtown Lexington (*below right*) and a bicycle identification program (*above right*) are parts of the community response to increased crime.

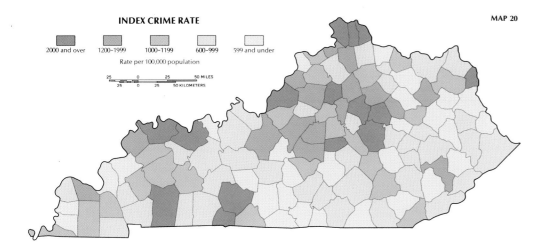

MAP 20

INDEX CRIME RATE

2000 and over 1200–1999 1000–1199 600–999 599 and under
Rate per 100,000 population

25 0 25 50 MILES
25 0 25 50 KILOMETERS

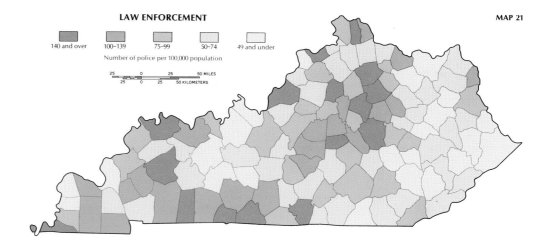

MAP 21

LAW ENFORCEMENT

140 and over 100–139 75–99 50–74 49 and under
Number of police per 100,000 population

25 0 25 50 MILES
25 0 25 50 KILOMETERS

CRIME AND LAW ENFORCEMENT

MAPS 20 and 21

Crime is one of the most serious problems confronting our society today. **Maps 20 and 21** present two distinct aspects of the geography of crime.

The distribution of index crimes in 1973, compiled from Kentucky State Police reports, is shown on **Map 20.** Seven types of crime are included: murder and manslaughter, forcible rape, robbery, aggravated assault, breaking and entering, larceny, and auto theft. These crimes are used as an index of total criminal occurrence because they are considered to be serious and are well reported. "White collar crimes," such as embezzlement and forgery, and "victimless crimes," such as prostitution or drunkenness, are not included because of poor reporting.

The crime rates mapped here are calculated per 100,000 population. This may lead to some distortion because most of the crimes included are directed against property. Wealthier counties have more property at risk per capita; thus, crime rates based on population may exaggerate the risk of crime in these wealthier counties. Furthermore, some counties with very small populations may have rather widely fluctuating rates from year to year.

Only a few counties reported crime rates above the state average of 2,273 per 100,000 population in 1973. Most of the high-rate counties were located in metropolitan areas. Relatively high rates were found also in rural counties of the Bluegrass. Lowest reported rates were found in rural counties of the Mountains and in the south-central part of the state.

The number of police department employees per 100,000 population (**Map 21**) is an indication of the effort made to reduce crime within various local areas. All uniformed and civilian employees of county sheriffs and of county and municipal police departments are included in calculating police employment. Campus police at the University of Kentucky and Eastern Kentucky University are also included. Not surprisingly, the largest numbers of law enforcement personnel are found in counties with the highest crime rates. The most urban or metropolitan counties have the largest numbers of police per 100,000 population, with Fayette County, at 321, in the lead. Rural counties in which there are no municipal police departments have the smallest proportions.

PHILLIP D. PHILLIPS

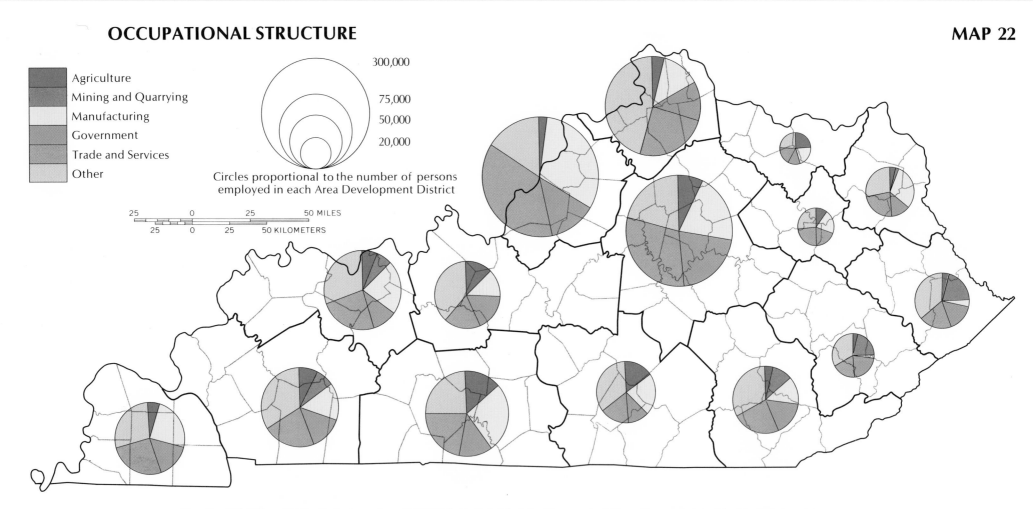

Agriculture
Mining and Quarrying
Manufacturing
Government
Trade and Services
Other

300,000
75,000
50,000
20,000

Circles proportional to the number of persons
employed in each Area Development District

25 0 25 50 MILES
25 0 25 50 KILOMETERS

MAPS 22-25

OCCUPATIONAL STRUCTURE AND EMPLOYMENT

The occupational structure of Kentucky's labor force (**Map 22**) reflects both the regionality of the state and the transition from a predominantly agricultural to a predominantly manufacturing society. Since World War II, growth in secondary and tertiary industries has been rapid. Approximately 250,000 nonagricultural wage and salary jobs were added during the 1960–1970 period. In 1974 employment in Kentucky totalled approximately 1,347,000 persons. (See **Figure 1**.) Recent occupational changes are shown in **Table 4**.

The greatest numbers of manufacturing workers are in those Area Development Districts with large urban centers. Most workers employed in trade (wholesale and retail) are likewise in the ADDs

having larger numbers of towns and cities. In the easternmost part of the state, where coal mining plays a major role, little manufacturing employment is found except in the northeast, and few persons make a living from agriculture. Larger percentages of agricultural workers are located in the Bluegrass, the Pennyroyal, and the Western Coal Field. The small percentage of persons employed in mining in the Western Coal Field (Green River and Pennyrile Development Districts) reflects mining methods rather than production. The region produces half of Kentucky's coal but most of it is produced by surface mining, which requires fewer employees than do underground operations. The percentages of persons employed in trade and services and in government do not vary greatly over

Figure 1. Occupational Structure, 1974
(Employment in 1,000s)

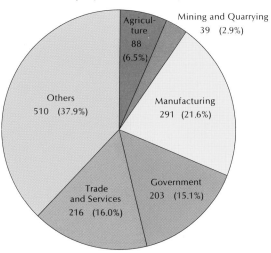

Total employment: 1,347,000

TABLE 4
Changes in Occupational Structure

Occupation	Employment (in 1,000 s)		Percentage of change
	1970	1974	
Agriculture	95	88	− 7.4
Mining and quarrying	32	39	+21.9
Manufacturing	265	291	+10.2
Government	190	203	+ 7.0
Trade and services	200	216	+ 8.0
Others	466	510	+ 9.4
Total employment	1,248	1,347	+ 8.0

the state. The diversity of industries in Kentucky insures stability in employment and an unemployment rate for the state as a whole that is normally less than the national rate.

Man's economic activities may be divided into three basic categories:

(1) Primary activities, which involve the direct exploitation of the earth and provide man with his basic sustenance: fishing, forestry, farming, and mining. The major primary activities in Kentucky revolve around livestock and crop production and the mining of coal.

(2) Secondary activities, which process the commodities gained from primary activities into useful products. Kentucky's major secondary activities are manufacturing of primary metals, metal fabrication, manufacturing of electrical and nonelectrical equipment, and food processing.

(3) Tertiary activities, in which persons are paid for the service they perform rather than the goods they produce: wholesaling, retailing, and maintenance of goods; finance and commerce; government employment; and construction and transportation. Most of the labor force of Kentucky in tertiary positions consists of professional and clerical workers and persons involved in wholesale and retail trade.

The character of employment in these three sectors within Kentucky's economy is similar to the national pattern. The Common-

wealth's labor force in 1970 was divided as follows: primary sector, 9.3 percent (**Map 23**); secondary sector, 25.6 percent (**Map 24**); tertiary sector, 65.1 percent (**Map 25**).

One should bear in mind the uneven distribution of Kentucky's labor force. Only seven counties—Campbell, Daviess, Fayette, Jefferson, Kenton, McCracken, and Warren—had labor forces in excess of 20,000 each in 1970, but they account for 45 percent of the state's labor force. Conversely, the sixty-eight counties with labor forces under 5,000 each comprise only 18 percent of the total labor force.

Only Metcalfe County has a plurality of its labor force in primary activities, although such employment is significant in most counties. In ninety-three counties 10 percent or more of the labor force is employed in primary activities. In forty-two of these counties at least 20 percent is in primary activities. While the primary production of many of these counties may not be of great importance to the overall productivity of Kentucky, it is locally significant as a source of employment and income.

It is in eastern Kentucky that primary employment forms the greatest proportion of the labor force. Mining, farming, and forestry are of crucial importance, although their role in the economy varies considerably within the region. The proportion of the labor force in mining generally decreases from east to west, while the proportion in farming and forestry increases. In both the Outer Bluegrass and the eastern Pennyroyal farming is by far the most significant primary activity. Mining and agriculture are exceptionally important in the Western Coal Field, the former being the dominant source of employment in Hopkins and Muhlenberg counties. As **Table 4** shows, employment in agriculture is declining in the state.

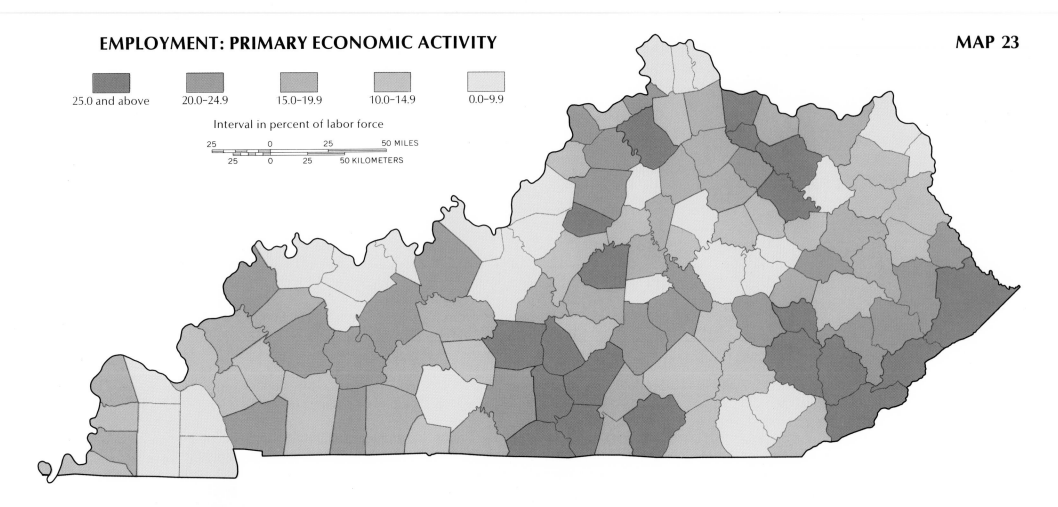

25.0 and above   20.0–24.9   15.0–19.9   10.0–14.9   0.0–9.9

Interval in percent of labor force

25   0   25   50 MILES
25   0   25   50 KILOMETERS

A surface coal-mining operation near Manchester in Clay County. Out of 39,293 persons employed in the Kentucky coal-mining industry in 1974, surface mining employed 15,014, although more than half the coal is surface mined. From 1973 to 1974 employment in this primary industry increased by 8,732 persons.

Equally clear in **Map 23** is the relative insignificance of primary activities in the most metropolitan counties. This results from the diversity of urban-oriented land uses and the more lucrative and steady income opportunities that exist in other sectors of the economy.

Kentucky is strategically located between the American industrial belt to the north and the increasingly industrial South. The growing affluence and markets in both areas have been key factors in the expansion of Kentucky's manufacturing since World War II **(Map 24)**. Also contributing to this expansion are the availability of power at competitive rates, a sizable labor force, a variety of tax concessions, and rapidly improving accessibility as the expressway network has expanded. The textile/apparel, tobacco, and bourbon industries have declined relatively, as measured by proportions of employment, while heavy industries, notably chemicals, primary

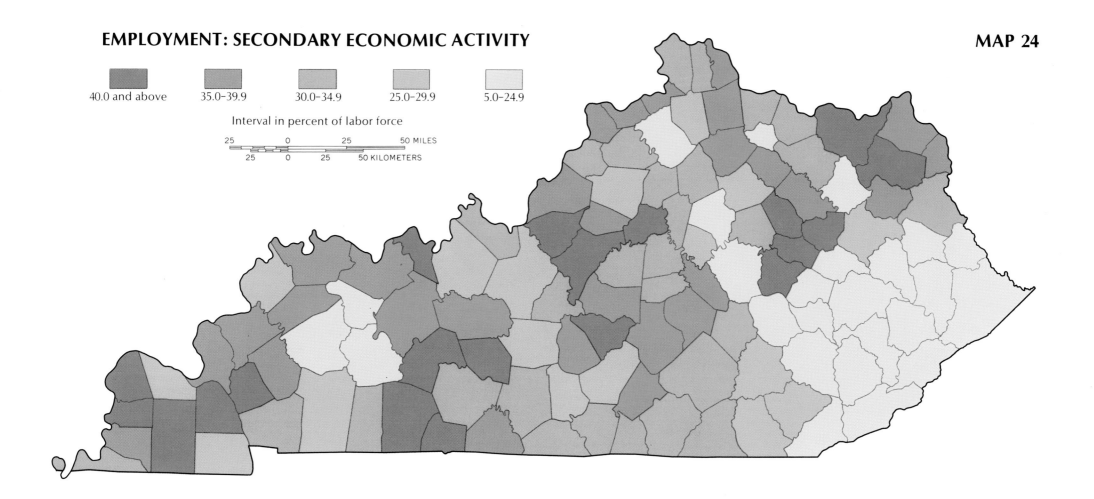

40.0 and above   35.0-39.9   30.0-34.9   25.0-29.9   5.0-24.9

Interval in percent of labor force

25   0   25   50 MILES
25   0   25   50 KILOMETERS

metals, and metal fabrication, have grown rapidly. Industrial diversity has been further increased by the expansion of the electronics industry. (**Maps 62–65** provide more detailed information on manufacturing.)

Much expansion of secondary activities has occurred outside Kentucky's traditional industrial nodes. The growth of metals industries in Henderson and Hancock counties is one example. **Map 24** must be interpreted with caution. Areas which appear to be most heavily industrialized actually account for only a small portion of the Commonwealth's total manufacturing employment and production. Most of these areas have small populations and limited total employment. The high proportions of employment in secondary activities conceal a weak or undeveloped tertiary sector.

Despite recent trends involving spatial decentralization of manufacturing, eight counties with sizable urban centers—Boyd, Campbell, Daviess, Fayette, Jefferson, Kenton, McCracken, and Warren—accounted for over 53 percent of Kentucky's 1970 manufacturing labor force and contained over half of the factories with 100 or more employees. The industrial expansion of the Inner Bluegrass and central Pennyroyal is noteworthy. Counties along the Ohio River, however, continue to account for nearly two-thirds of the state's manufacturing as measured by value added and payrolls. Despite efforts by the Appalachian Regional Commission, eastern Kentucky has almost no significant manufacturing employment. Few counties in this region maintain a secondary labor force equal to or in excess of the 25.6 percent average of Kentucky as a whole.

*Occupational Structure
& Employment*   41

MAP 25

EMPLOYMENT: TERTIARY ECONOMIC ACTIVITY

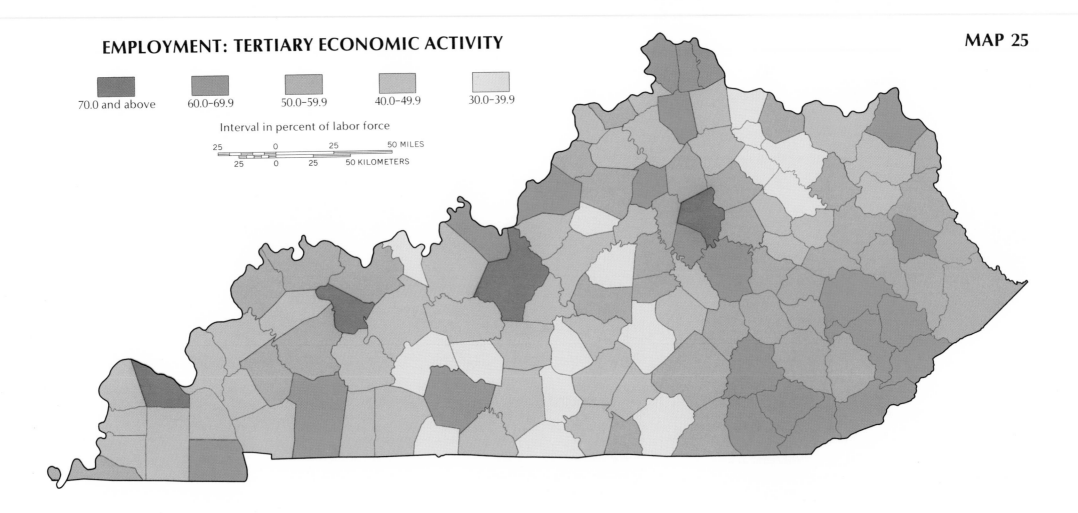

70.0 and above 60.0-69.9 50.0-59.9 40.0-49.9 30.0-39.9

Interval in percent of labor force

25   0   25   50 MILES
25   0   25   50 KILOMETERS

With the exception of Metcalfe County, all of Kentucky's counties have a plurality of their labor forces in tertiary activities (**Map 25**). In only sixteen counties does this sector employ less than 40 percent of the labor force. The character of tertiary employment varies considerably among the counties, however. Hardin County, for example, reflects the service center character of Elizabethtown, located at the crossroads of two major expressways, as well as employment opportunities for civilians at military installations. In addition nearly 14 percent of the 1970 Hardin County labor force was employed in public administration. In Franklin County, containing the state capital, government employment exceeds 20 percent. By contrast, 25 percent of Breathitt County's labor force was employed in private or public education in 1970. Educational employment is also significant in Fayette, Rowan, Madison, and Calloway counties, sites of large state universities. Throughout the Mountains employment in the public sector and in the construction industry is relatively high. The work opportunities afforded by the expansion of the highway network in eastern Kentucky are reflected here.

In terms of total employment in the tertiary sector, however, the large urban centers are dominant. Jefferson County annually accounts for approximately 25 percent of the retail trade and 50 percent of the wholesale trade of Kentucky. (See **Maps 66 and 67**.) The variety of services and the size of the labor force involved in tertiary activities clearly demonstrate the service character of most large Kentucky settlements.

WILFORD A. BLADEN, TERRY L. McINTOSH, and WILLIAM A. WITHINGTON

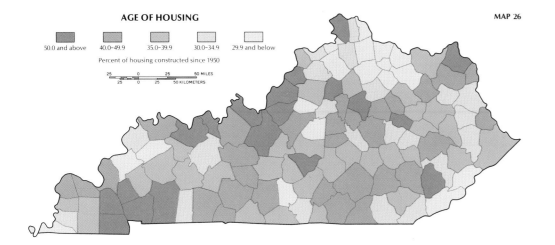

MAP 26

AGE OF HOUSING

50.0 and above 40.0-49.9 35.0-39.9 30.0-34.9 29.9 and below

Percent of housing constructed since 1950

25 0 25 50 MILES
25 0 25 50 KILOMETERS

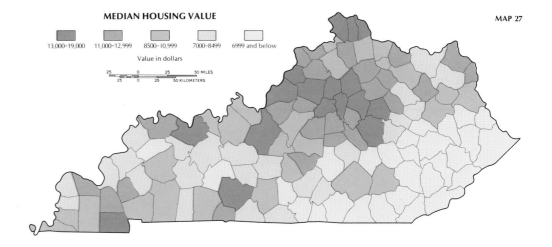

MAP 27

MEDIAN HOUSING VALUE

13,000-19,000 11,000-12,999 8500-10,999 7000-8499 6999 and below

Value in dollars

25 0 25 50 MILES
25 0 25 50 KILOMETERS

HOUSING CHARACTERISTICS

MAPS 26-29

Age of housing (**Map 26**) is an indicator of both quality and style of housing. Newer housing is generally in better structural condition than older housing, is more likely to have complete plumbing facilities, and is often better adapted to today's smaller families. Housing constructed since 1950 tends to be found in greater quantity in the metropolitan areas of the Commonwealth. A major exception is the Covington-Newport area, where large numbers of older housing units within the inner cities outweigh newer suburban housing. Much of the housing in Trigg, Lyon, and Marshall counties in western Kentucky is also newer, probably because of displacement of persons by creation of Lakes Barkley and Kentucky and the development of Land Between the Lakes. The largest proportion of older housing is in the Outer Bluegrass between Lexington and Covington-Newport, historically one of the slowest growing areas in the state. Maysville and Washington in Mason County are well known for their fine old homes, many of which are in the National Register of Historic Places. Other areas of older housing are scattered throughout the state. In general the Mountain areas have newer housing than might be anticipated from population

change and housing quality. Here, however, much of the newer housing is small and often poor in original construction quality.

The median values of housing as estimated by the owners (**Map 27**) reflect both the quality of housing and the demand for it. Areas with lower value housing tend to have smaller, older housing units, more of which lack piped water, flush toilets, sanitary sewers, and other amenities. Even more important is the general local demand for housing. The areas of lowest housing value are all areas of population decline or very slow growth. A house in these areas will have a much lower value than a comparable house in a rapidly growing area. As a result of variations in quality and demand, the median value of housing ranges fourfold from under $5,000 in some counties to a high of about $19,000 in Jefferson County. The Mountains, southeastern Pennyroyal, and portions of the Western Coal Field have the lowest median housing values. Highest values are in the Inner Bluegrass and the larger urban areas.

The percentage of housing units which have complete plumbing facilities (**Map 28**) is a good indicator of general housing quality. A unit must have hot and cold piped water within the structure

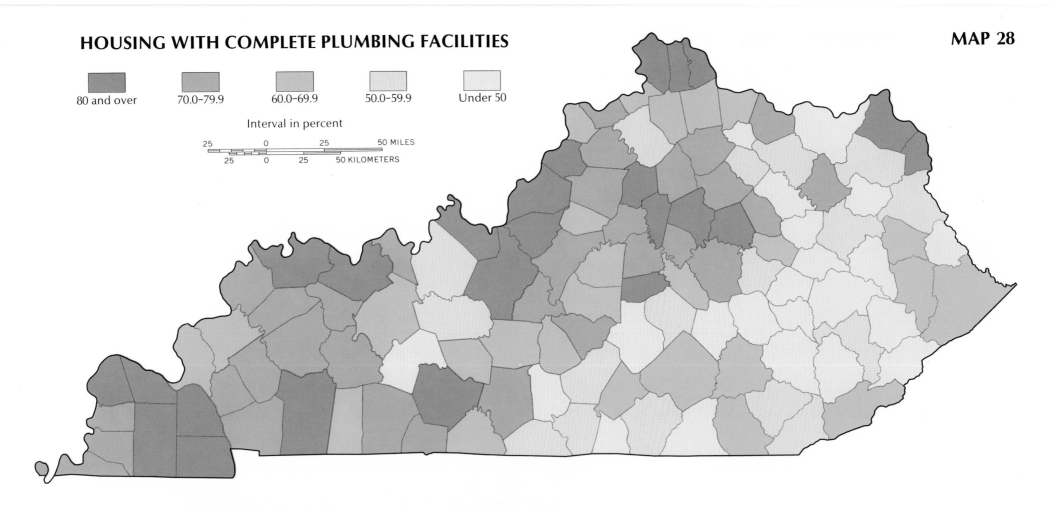

80 and over | 70.0–79.9 | 60.0–69.9 | 50.0–59.9 | Under 50

Interval in percent

and a flush toilet and bathtub or shower for the exclusive use of the occupants in order to be classified by the Census Bureau as having "all plumbing facilities." A high correlation exists between units with complete plumbing facilities and units with complete kitchen facilities and uncrowded living conditions (less than one person per room). Therefore, complete plumbing facilities have been used as a surrogate for general housing quality.

The highest percentages of housing units with complete plumbing facilities are found in metropolitan counties. Only three (Bourbon, Scott, and Jessamine) of the seventeen counties defined by the Census Bureau in 1974 as parts of Standard Metropolitan Statistical Areas have less than 80 percent of all housing units with complete plumbing facilities. The highest levels are also found in other major urban areas. The lowest levels, under 50 percent, are found in the Mountains and parts of the Western Coal Field.

HOUSING ATTACHED TO A PUBLIC SEWER

MAP 29

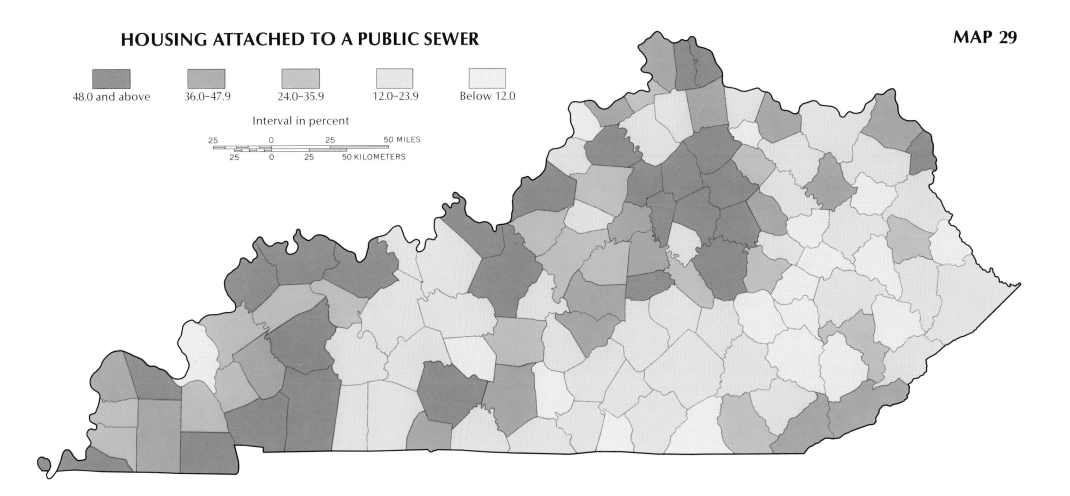

48.0 and above   36.0-47.9   24.0-35.9   12.0-23.9   Below 12.0

Interval in percent

25   0   25   50 MILES
25   0   25   50 KILOMETERS

Housing units attached to public sewers **(Map 29)** are concentrated in the metropolitan counties of the state, particularly within the cities. Many if not most suburban residences in the large metropolitan areas have either septic tanks or small private "package" sewer plants. The tremendous amounts of sewage generated and the relatively high population densities have led to serious health problems in some of these areas, and provision of adequate sewage treatment has become a major political issue. Rural counties of the Commonwealth have uniformly low levels of public sewer provision. The seriousness of the lack of public sewers, however, depends greatly on the type and amount of soil, type and depth of subsoil, and type and closeness of bedrock to the surface, all of which vary widely.

PHILLIP D. PHILLIPS

The mountaineer's home without plumbing (*opposite*) is characteristic of depressed areas in eastern Kentucky. In contrast, new homes like these in an Owensboro subdivision have complete modern plumbing. Some subdivisions built shortly after World War II, however, lacked adequate sewage treatment facilities and other public services, creating problems only now being dealt with by the cities.

Housing 45

V. EDUCATIONAL DEVELOPMENT

MAPS 30-32

Although the maps focusing on educational characteristics do not specifically emphasize the point, the availability and structure of education for Kentuckians have greatly improved in the past two decades. Where one-room schools once were the standard form of community education, few are now used. Instead, consolidated schools drawing on larger student bodies and capable of stronger educational involvement are increasingly typical, except in the larger urban areas, where many schools are needed. At junior and senior high school levels, county consolidated schools have taken over in most areas where local urban communities are not large enough to support strong schools. At higher education levels, Kentucky today has a geographically extensive system of eight state-supported universities. Fourteen state-supported community colleges, administered as a group from offices within the University of Kentucky at Lexington, provide introductory and specialized two-year college programs in many intermediate and smaller urban centers across the state.

Throughout its history Kentucky has had many small private institutions of higher education with strong traditions of learning. Transylvania University in Lexington is a notable example, having had a university-level medical school of high reputation in the 1820s and 1830s. A great many other private colleges with religious affiliations or without specific associations are found in communities throughout Kentucky. Berea College, looking to service in the southeastern hills and mountains from its site at the juncture of the Bluegrass and encircling Knobs, is a highly regarded example of a small school standing for high-quality education in a self-supporting program of student work.

This modern elementary classroom reflects educational progress in Kentucky, which has lagged behind the rest of the nation. Many communities are seeking improvements in education.

MAP 30

MEDIAN SCHOOL YEARS COMPLETED

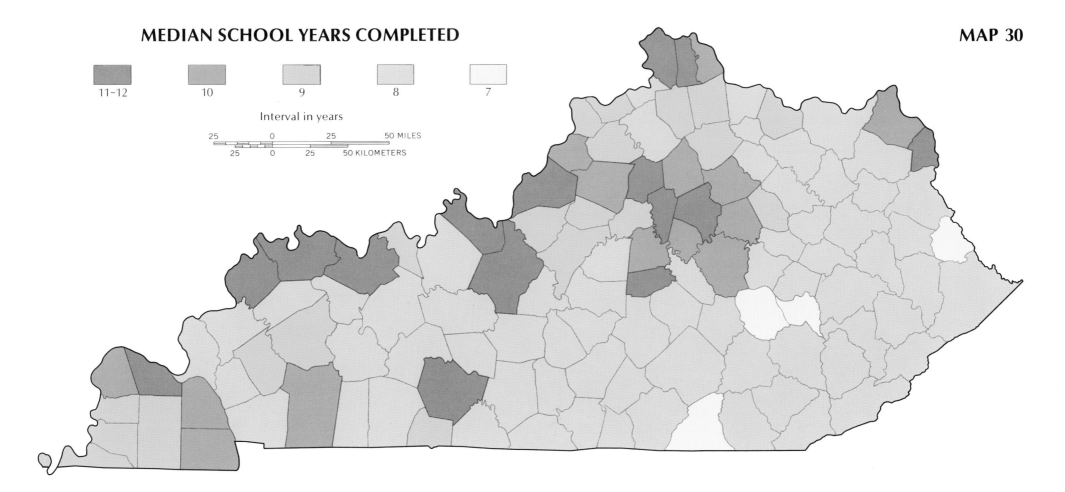

11–12 10 9 8 7

Interval in years

25   0   25   50 MILES

25   0   25   50 KILOMETERS

Map 30, Median School Years Completed, like the other two maps in this chapter, reflects most clearly the pattern of counties in 1970 having a combination of larger urban communities and relatively high per capita income. Twelve of the fourteen counties in which median school years completed was eleven or twelve had urban centers of at least 10,000. One exception, Union County, lies immediately west of Henderson, with per capita income and percentage of gain in per capita income somewhat below Kentucky's averages. The other exception, Woodford County, is rapidly growing as part of the urbanizing central Bluegrass.

Most of the state's counties have median education levels of eight years or less. A broad swath of Kentucky is included, from the Cumberland River on the west across west-central and south-central parts of the state into eastern Kentucky and some of the hillier parts of northern Kentucky south of the Cincinnati-Coving-

ton metropolitan area. Only Boyd and Greenup counties in the Ashland-Huntington metropolitan area are positive exceptions.

The percentage of Kentuckians with no high school education dropped from 70.4 in 1940 to 43.2 in 1970. The percentage with no formal education also dropped, from 4.1 to 1.7 during the same period. The comparison with national figures is less favorable, however. The 1940 national educational median was 8.4 years, compared with 8.2 years for Kentucky. But by 1970 the national median had jumped to 12.2 years while Kentucky had a 1970 median of 9.9 years and ranked fiftieth among the states. The 1970 census shows that only 38.5 percent of Kentuckians twenty-five years and older have completed high school or some college, as compared to the national average of 52.3 percent. Kentucky is tied with North Carolina for forty-eighth place in this category. Only South Carolina and Arkansas rank lower.

MAP 31

HIGHER EDUCATION

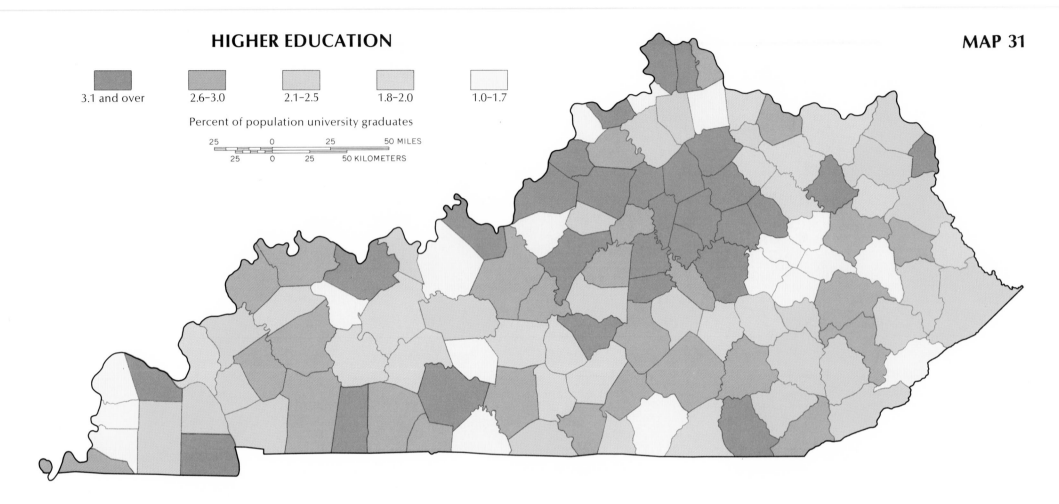

| 3.1 and over | 2.6–3.0 | 2.1–2.5 | 1.8–2.0 | 1.0–1.7 |

Percent of population university graduates

25   0   25   50 MILES
25   0   25   50 KILOMETERS

A dramatic rise has occurred since 1955 in the numbers of students enrolled in public and private institutions of higher learning. Also noteworthy is the rapid increase in the proportion of persons in ages seventeen to twenty-five enrolled in or having already completed some form of college or university-level education.

The proportion of college graduates to total population in Kentucky is approximately 4.44 percent. The proportion of college graduates by county (**Map 31**) forms a pattern similar to the one for median school years completed, but is more complex. The counties which are most highly urbanized and which also have the highest per capita incomes are usually the ones with the highest proportions of college graduates.

The largest zone of counties with more than 3 percent of their population college graduates focuses on Fayette and Jefferson counties, including the intervening and surrounding Outer Blue-

grass counties. This area contains a high concentration of large state-supported universities, smaller private colleges, and relatively high proportions of graduates of these and other institutions. The University of Kentucky in Lexington, Eastern Kentucky University in Richmond, the University of Louisville, and Kentucky State University in Frankfort are the large state-supported university-level schools; Georgetown, Transylvania, Berea, and Centre are among the many smaller colleges in this zone.

A second, much smaller cluster in northern Kentucky includes Boone and Kenton counties in Cincinnati's metropolitan area and Carroll County to the southwest. Northern Kentucky University is the principal state-supported institution in the area. Other college-level institutions easily accessible by good highways lie northward across the Ohio River in Cincinnati, southwestward in Louisville, and southward in the Bluegrass.

Murray State University (*far left*), educational center for the Jackson Purchase, is an example of Kentucky's efforts to keep pace with other states in providing higher public education in all regions of the state.

Berea College is internationally renowned for its tuition-free work-study program for Appalachian students. The program focuses on traditional crafts, such as hand weaving, demonstrated at left.

Widely scattered single-county areas with high proportions of college graduates include most of Kentucky's other large urban centers having community colleges, four-year colleges, or universities. Among these are Rowan County (Morehead State University); Whitley County (Cumberland College); Warren County (Western Kentucky University); Calloway County (Murray State University); and Boyd, McCracken, and Daviess counties, each of which has a community college. Meade, Todd, and Nelson counties have high proportions of college graduates since each is near one or more sizable urban centers having colleges or universities.

The counties with lowest proportions of college graduates are widely distributed across the state. One cluster is located along the Mississippi River in the Jackson Purchase and another is centered in the Mountains.

In 1973, the eight state-supported Kentucky universities had a total enrollment of nearly 78,000 students, more than 50 percent higher than enrollment five years earlier (**Table 5**). Two institutions, Northern Kentucky State College (now Northern Kentucky University) and the University of Louisville, had been added to the system. The University of Kentucky, with more than 21,000 students in 1973, was nearly twice as large as any of the next three. About 85 percent of all students in 1973 were Kentucky residents, compared to about 77 percent in 1968.

TABLE 5
Student Enrollment at State-Supported Universities

	Fall 1968 enrollment			Fall 1973 enrollment		
Institution	Kentucky students	Other students	Total enrollment	Kentucky students	Other students	Total enrollment
Univ. of Kentucky[a]	11,798	3,472	15,270	17,602	3,794	21,396
Western Kentucky Univ.	9,119	1,451	10,570	11,138	1,192	12,330
Univ. of Louisville[b]	–	–	–	10,738	1,492	12,230
Eastern Kentucky Univ.	7,164	2,016	9,180	9,156	1,932	11,088
Murray State Univ.	5,109	2,225	7,334	6,248	1,279	7,527
Morehead State Univ.	4,575	1,687	6,262	5,254	1,324	6,578
Northern Kentucky Univ.[c]	–	–	–	4,176	582	4,758
Kentucky State Univ.	1,105	505	1,610	1,544	456	2,000
Total	33,870	11,356	50,226	65,856	12,051	77,907
Percentage	77.4	22.6	100.0	84.54	15.46	100.0

[a]Enrollment at community colleges not included.
[b]Not in system in 1968.   [c]Not established in 1968.

Maps of university or college-level institutions usually express data in terms of numbers of students attending the colleges or universities. **Map 32,** in contrast, shows the origin of in-state student enrollments at Kentucky's eight state-supported universities and colleges. The circles are proportional to the number of students from each county attending these institutions. The pattern within each "pie diagram" indicates the proportion of students from the county attending particular ones among the eight institutions.

Three counties stand out as major source areas of students, and several others as of intermediate importance. The three principal areas are Jefferson County, the single largest source area; the central Bluegrass, principally Fayette County but including also Franklin and Madison counties as well as five counties adjoining Fayette; and northern Kentucky, including Kenton, Campbell, and Boone counties. Intermediate source areas are Warren County in the Pennyroyal; Daviess County on the Ohio River; a series of five counties from Christian and Trigg into the Jackson Purchase counties of Calloway, Graves, and McCracken; and finally, four counties in eastern Kentucky: Boyd, Greenup, Carter, and Rowan.

Many counties of southeastern, southern, and central Kentucky also provide large numbers of students to various institutions, no one of which predominates.

Student choices tend to be dominated by the nearest or second nearest state institution. For example, in northeastern Kentucky most of the counties contribute their largest proportions of students to Morehead State University in Rowan County. In the southeastern area Eastern Kentucky University dominates in many counties; in many others a three-way division between Eastern Kentucky University, Morehead State University, and the University of Kentucky is characteristic, as in Pike, Floyd, and Letcher counties. The University of Kentucky draws most heavily on Fayette and several nearby counties, but as the principal state university it also draws on many counties which might be expected to have other affiliations.

WILLIAM A. WITHINGTON and WILFORD A. BLADEN

ORIGIN OF UNIVERSITY STUDENTS

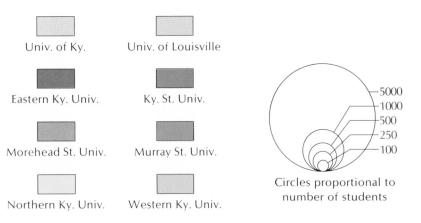

Univ. of Ky.   Univ. of Louisville

Eastern Ky. Univ.   Ky. St. Univ.

Morehead St. Univ.   Murray St. Univ.

Northern Ky. Univ.   Western Ky. Univ.

5000
1000
500
250
100

Circles proportional to
number of students

For counties with 500 or fewer students, the
circle shows only the predominant school

20   0   20   40 MILES

20   0   20   40 KILOMETERS

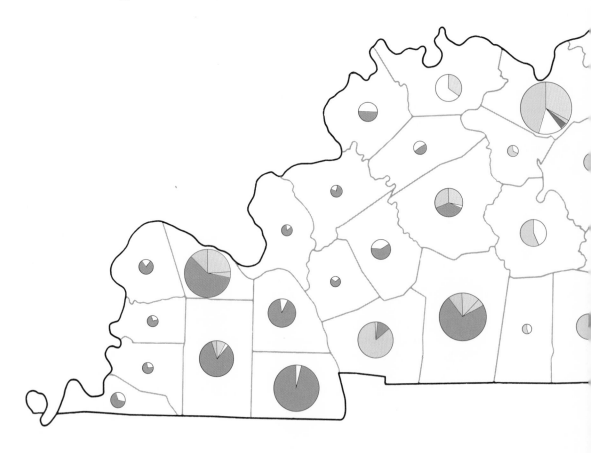

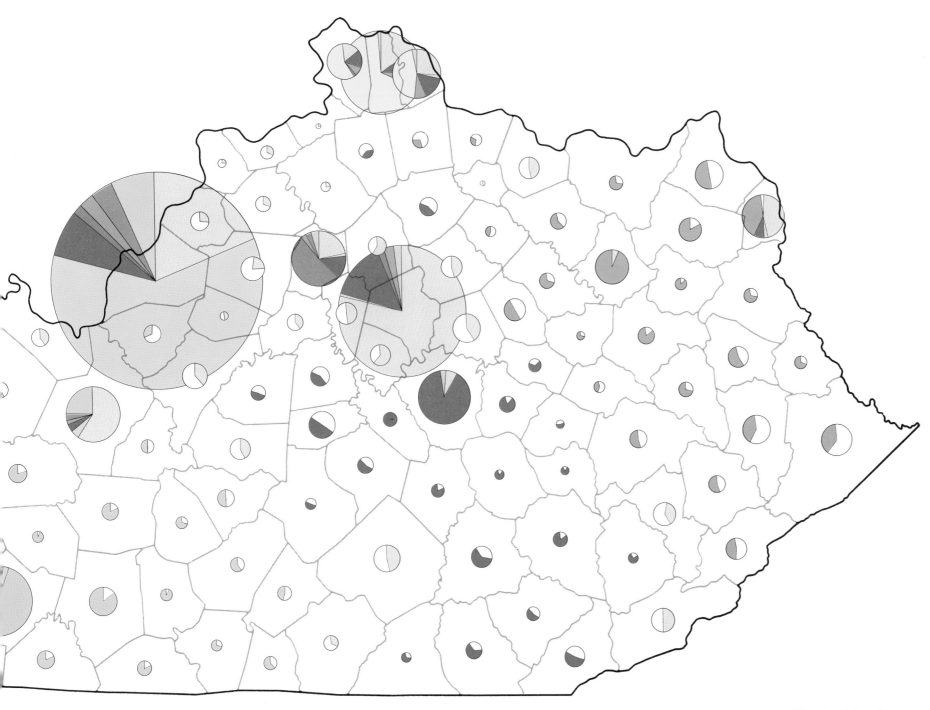

MAP 32

VI. PATTERNS OF HEALTH CARE

MAPS 33-37

Two serious health-care problems faced by Kentuckians are the low numbers of health-care personnel and their uneven distribution. Of the fifty states, Kentucky ranked forty-second in the rate of physicians and forty-third in the rate of dentists per 100,000 population in 1971. In 1966 Kentucky ranked forty-second in the rate of nurses per 100,000 population. These rates were about 60 percent of the national rates for physicians and nurses and about 75 percent of the national rate for dentists.

While low numbers of health-care personnel are a problem for the state as a whole, the problem is aggravated further in certain areas by their uneven distribution. The numbers of persons per physician, dentist, and nurse in the largest cities compare favorably with United States averages, but several counties are far below the national average. The problems created by uneven distribution of personnel are worsened in areas where people are poor, roads are poor, and distances to cities are great. Urban counties with the lowest numbers of persons per physician **(Map 33)** include Jefferson (Louisville), Fayette (Lexington), Boyd (Ashland), Franklin

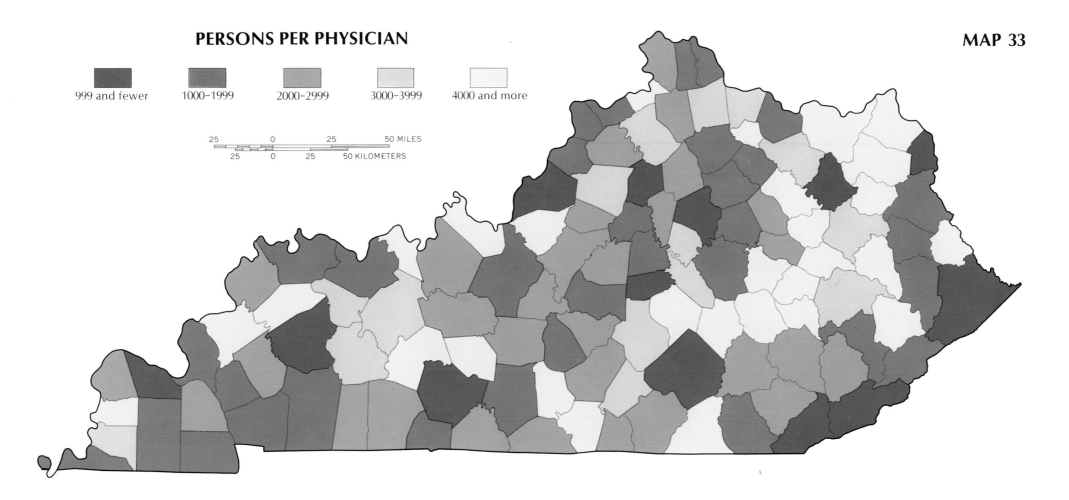

999 and fewer 1000–1999 2000–2999 3000–3999 4000 and more

25   0   25   50 MILES
25   0   25   50 KILOMETERS

(Frankfort), Warren (Bowling Green), and McCracken (Paducah). This phenomenon is expected in part because of the location of specialists in larger cities and in part because of the financial and cultural attractiveness of cities to physicians.

A generally recognized benchmark is the figure of 3,000 persons per physician, above which a critical shortage is said to exist. In 1975 forty-four counties in Kentucky had over 3,000 persons per physician. In some instances this figure is misleading; in other cases it indicates a serious problem. While people in Bullitt County, a bedroom county for Louisville, most likely get physicians' services in Louisville, people in Knott and Magoffin counties are far from places with adequate numbers of physicians. Keeping this in mind, the most serious shortage areas for physicians within the state are in the counties of the Mountains to the east and southeast of the Bluegrass.

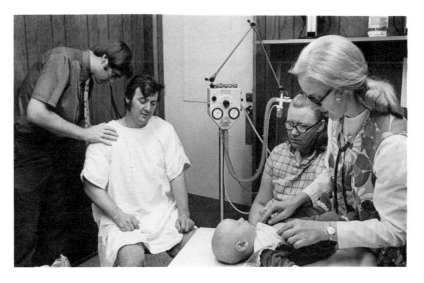

Mary Breckinridge Hospital at Hyden (*opposite*), built in 1928, represents a major effort to bring modern health services to eastern Kentucky. The area has suffered from a chronic shortage of health-care personnel. Albert B. Chandler Medical Center in Lexington also operates health-maintenance clinics (*left*) at various places in the Mountains.

PERSONS PER DENTIST

MAP 34

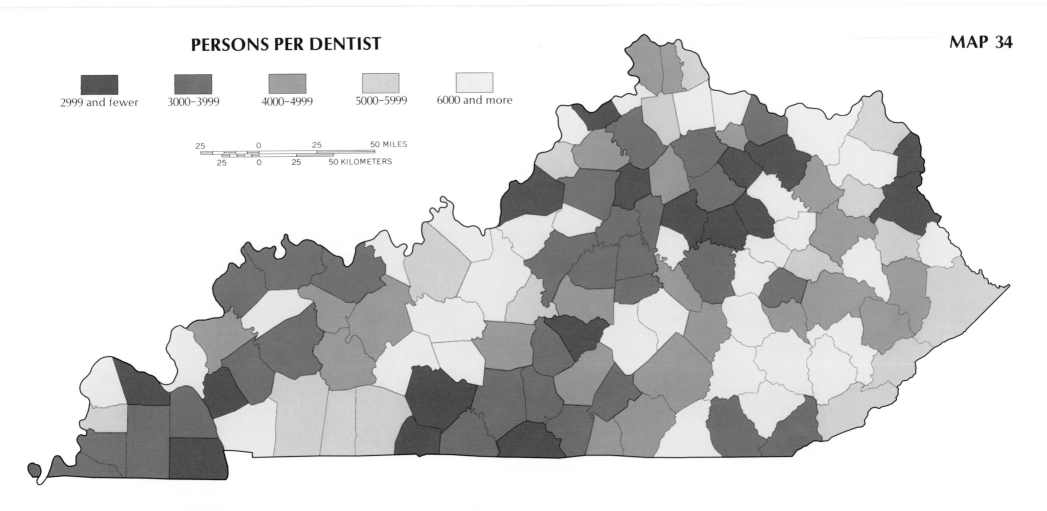

2999 and fewer   3000–3999   4000–4999   5000–5999   6000 and more

25   0   25   50 MILES

25   0   25   50 KILOMETERS

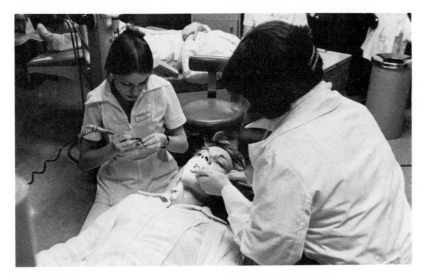

These dental technicians being trained in a Louisville hospital will help reduce the shortage of dental personnel in Kentucky. In most communities better nutrition and health education are improving the dental condition of Kentuckians.

The geographical distribution of persons per dentist **(Map 34)** is similar to that of physicians. The lowest numbers, indicating the largest numbers of dentists relative to population, are in the larger cities. Once again McCracken, Warren, Jefferson, Franklin, Fayette, and Boyd counties stand out. The national average for the number of persons per dentist is approximately 2,200 to 1. A critical shortage, as defined by the U.S. Department of Health, Education, and Welfare, is 5,000 or more persons per dentist. Using this standard, fifty Kentucky counties have critical shortages of dentists. The shortage areas are widespread but, as with physician data, the statistics must be interpreted with care. While Campbell and Kenton counties have relatively high populations per dentist, many people in this area receive dental care in Cincinnati. For the more isolated parts of the state, especially in the east, the shortage is a serious problem.

MAP 35

PERSONS PER NURSE

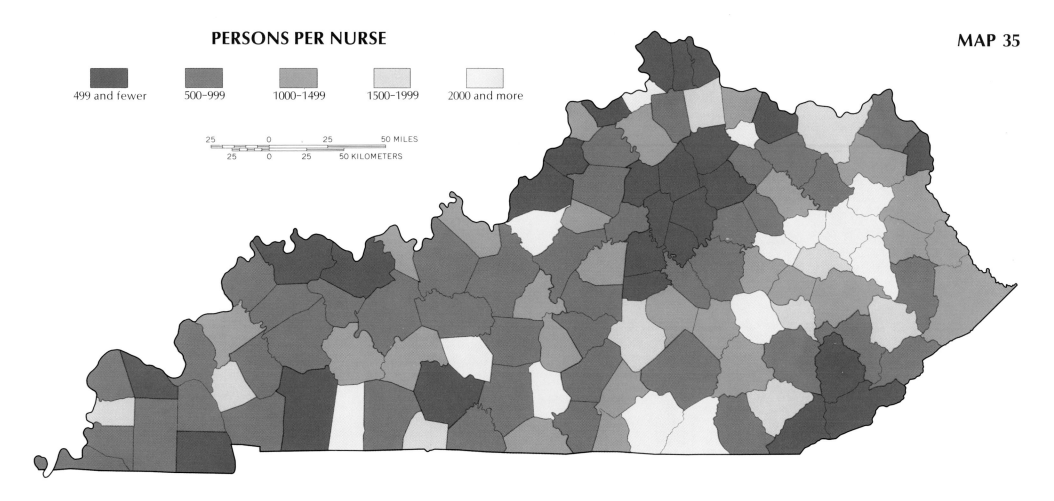

499 and fewer — 500-999 — 1000-1499 — 1500-1999 — 2000 and more

25 0 25 50 MILES
25 0 25 50 KILOMETERS

The number of persons per nurse in a county **(Map 35)** displays a similar pattern to those of physicians and dentists. Large cities in Kentucky have the lowest numbers of persons per nurse.

One of the unique features of health care in Kentucky is the Frontier Nursing Service, founded by Mary Breckinridge in 1925, which maintains nursing centers in remote areas of eastern Kentucky. Each center has a clinic and several nurse-midwives who provide general nursing and obstetrical service to an area of about eighty square miles. Before the service was founded, trained medical personnel were almost nonexistent in the area. Home remedies were widespread and maternal and infant death rates were high. Initially suspicious, the mountain people soon gained confidence in the nurses. Until recently the Mary Breckinridge Hospital at Hyden, built in 1928, was the only modern medical facility in the Mountain area.

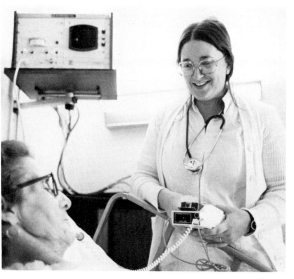

A patient receives modern medical care at Chandler Medical Center, Lexington. For long or unusual illnesses proper hospital and nursing care are essential but may be inaccessible in some sections of the state, as well as extremely costly.

PERSONS PER ACUTE BED

MAP 36

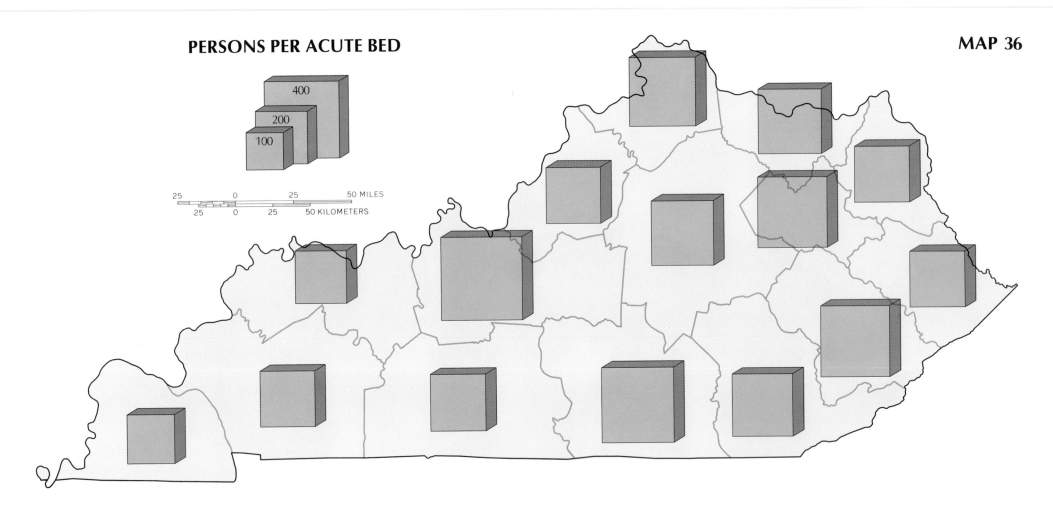

400
200
100

25   0   25   50 MILES
25   0   25   50 KILOMETERS

Medical research done at Chandler Medical Center, Lexington, is making possible new treatments for diseases such as cancer. Public health personnel work with the Medical Center staff in developing preventive health programs and better delivery of health services in the Commonwealth.

While Kentucky does not compare favorably with the United States as a whole in the numbers of persons per doctor, dentist, and nurse, the number of persons per acute bed is about the same as the national average, although the distribution is uneven (**Map 36**). Many medical problems can be handled effectively at small local hospitals, but very specialized services can be obtained only at one or two major facilities in the state, such as the large hospitals in the Bluegrass and KIPDA Area Development Districts. Thus the imbalances shown on **Map 36** do not represent a serious problem. Indeed, the situation for hospital facilities is much better than the situation for health manpower.

The infant death rate is measured as the number of infant deaths per 1,000 live births. **Map 37** shows infant deaths according to the mother's county of residence. Only births in hospitals are included.

MAP 37

AVERAGE INFANT DEATH RATE, 1968-1972

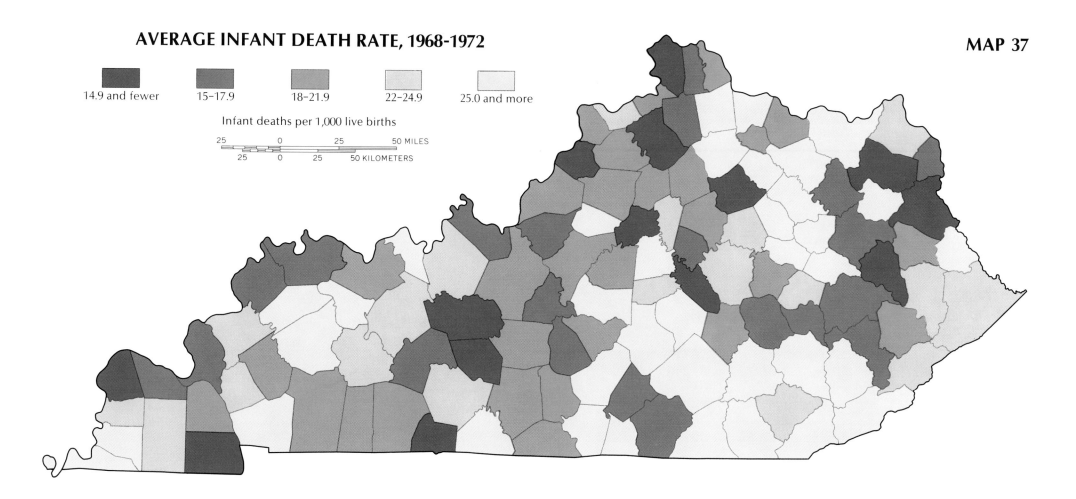

14.9 and fewer	15-17.9	18-21.9	22-24.9	25.0 and more

Infant deaths per 1,000 live births

25   0   25   50 MILES
25   0   25   50 KILOMETERS

An infant is considered to have "survived" if it leaves the hospital alive.

The average infant death rate for Kentucky for the five-year period covered here was about 20 per 1,000 live births. Rates of 25 or more generally exist in counties where incomes are below the state average and where health personnel are scarce. The counties containing Louisville and Lexington, the largest cities in the state, have relatively high infant death rates despite high incomes and major medical facilities, perhaps in part because of higher rates for blacks and large concentrations of blacks in these cities.

While the infant death rate for Kentucky as a whole remains higher than the national average, it is declining substantially, from 21.8 to 18.9 per 1,000 live births between 1968 and 1972.

RICHARD I. TOWBER

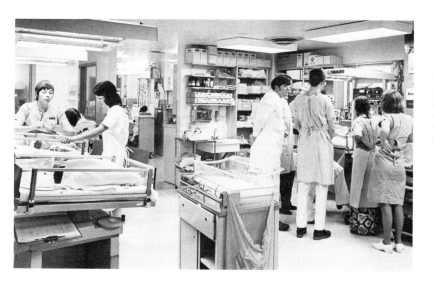

Emergency medical care for infants with birth defects in an intensive care ward such as this one at Chandler Medical Center, Lexington, has helped lower the state's infant death rate. All new babies receive immunizations which are reducing the incidence of communicable diseases in the state.

VII. TRANSPORTATION AND COMMUNICATION

MAPS 38-42

TRANSPORTATION NETWORK

In 1975 Kentucky was served by a railroad network of 3,762 route-miles **(Map 38)**. This network reaches most parts of the Commonwealth and provides intrastate and interstate linkages. The trunk-line routes extend primarily north–south, traversing Kentucky enroute from the industrial North and Midwest to most parts of the South. Secondary routes within Kentucky provide east–west connections focusing mainly on Louisville. East–west routes are conspicuous by their absence in southern Kentucky. The Appalachian coal fields are served by the Louisville and Nashville (L & N) and Chesapeake and Ohio (C & O) lines. A second concentration of coal routes, displaying a basically rectangular grid structure, is operated by the L & N and the Illinois Central Gulf (ICG) line in the less rugged Western Coal Field.

The rail network is dominated by freight traffic, which is the principal support of the system. Railroads are the backbone of coal transportation, especially in eastern Kentucky, where about 91 percent of the total volume of coal shipments to the consumer is by rail. Five railroads—the Louisville and Nashville, Chesapeake and Ohio, Illinois Central Gulf, Southern Railway, and Norfolk and Western—operate 95 percent of all rail-route mileage in Kentucky. These railroads account for the bulk of revenues and ton-miles, as well.

Three train routes have stops in Kentucky offering passenger service: the ICG route from Chicago to New Orleans, which stops at Fulton; the C & O route from Washington, D.C., to Chicago, which stops in Catlettsburg (previously in Ashland) and then goes

Rail yards at Paducah, a major trade and transportation center in the Jackson Purchase. With their ability to haul heavy loads cheaply, the railroads continue as the leading freight carrier in Kentucky.

MAP 38

MAJOR RAIL ROUTES

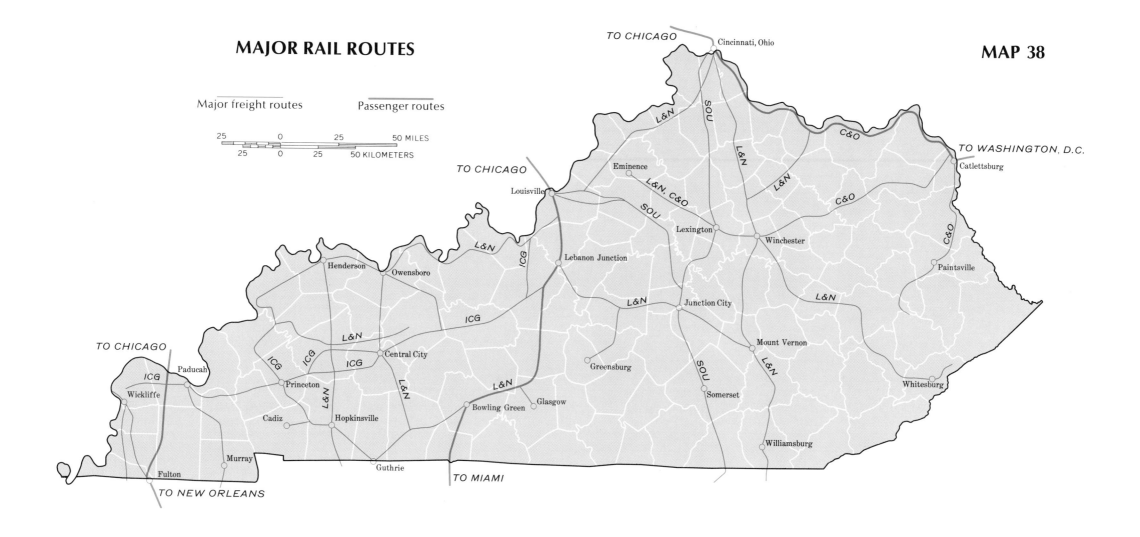

Major freight routes   Passenger routes

25   0   25   50 MILES
25   0   25   50 KILOMETERS

TO CHICAGO — Cincinnati, Ohio

TO WASHINGTON, D.C.

TO CHICAGO

Eminence

Louisville

L & N
SOU
C&O
L&N
C&O
L&N, C&O
Lexington
Winchester
C&O
Catlettsburg
Paintsville

Henderson
Owensboro
L&N
ICG
Lebanon Junction
SOU
L&N
Junction City
L&N
Mount Vernon

ICG
L&N
Central City
Greensburg
SOU
Somerset
Whitesburg

TO CHICAGO
ICG
ICG
ICG
Paducah
Wickliffe
Princeton
L&N
L&N
ICG
L&N
Bowling Green
Glasgow
Williamsburg

Cadiz
Hopkinsville

Murray
Fulton
Guthrie
TO MIAMI
TO NEW ORLEANS

through Kentucky to Cincinnati; and the L & N route from Chicago to Miami, which stops in Louisville and Bowling Green. There is a wide network of track which could be used for passenger service if economic, technical, and other factors should render it feasible. In 1974, a total of 33,281 passengers entrained or detrained in Kentucky, an average of 91 per calendar day. By city the totals were: Fulton, 3,677; Bowling Green, 4,395; Ashland, 5,895; and Louisville, 19,314. The Kentucky Legislative Research Commission recently carried out a feasibility study of east–west passenger rail service in the state. It concluded that at present such service is not feasible because of economic, technical, and institutional factors. The present passenger miles through Kentucky run operating deficits and a profit could not be expected for an east–west line. A further problem is that the existing east–west tracks are designed for freight trains and are not suitable for the operation of fast passenger trains. In addition, public concern for passenger rail service seems to be low. During the gasoline shortage of 1973–1974 Amtrak experienced a 30 percent increase in passengers, but by the summer of 1974 Americans appeared to be getting back into their cars and reverting to their former travel habits.

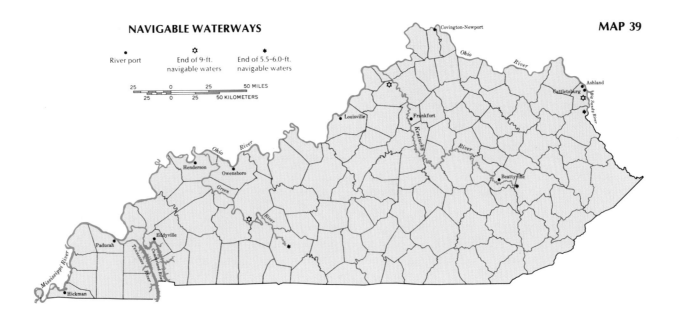

NAVIGABLE WATERWAYS

MAP 39

- River port
- ✿ End of 9-ft. navigable waters
- ✦ End of 5.5–6.0-ft. navigable waters

25   0   25   50 MILES
25   0   25   50 KILOMETERS

Kentucky is well served by navigable waterways. The extent of commercial waterways is shown on **Map 39.** Cities along the northern boundary of Kentucky have access to water transportation on the Ohio River. Although various commodities are transported on the Ohio, the principal commodities in and out of Louisville are coal, sand, gravel, crushed rock, and petroleum products. Western Kentucky is served by the Mississippi, Tennessee, and Cumberland rivers, as well as the Ohio. The principal commodities shipped on the Tennessee and Cumberland rivers are coal and nonmetallic minerals, while the Mississippi carries a wide variety of commodities.

In west-central Kentucky the Green River provides a six-foot channel from the Ohio River to the Mammoth Cave area, a distance of about 198 miles. A nine-foot channel extends from the Ohio River to about fifteen miles upriver from Central City, a distance of about 103 miles. The principal commodity on the Green River is coal moving from the Western Coal Field to the Ohio River.

Central Kentucky is served by the Kentucky River. Although the nine-foot channel extends only through Carroll County, the six-foot channel extends to a little beyond Beattyville, a distance of about 259 miles. At present, barge transportation generally does not extend beyond Frankfort. The principal commodities on the Kentucky River are sand, gravel, and crushed rock; some coal has been transported on the Kentucky in recent years, however.

A coal barge on Kentucky Lake. Inexpensive waterway transportation of bulk-type cargoes such as coal is an important factor in the Kentucky economy.

Transportation & Communication

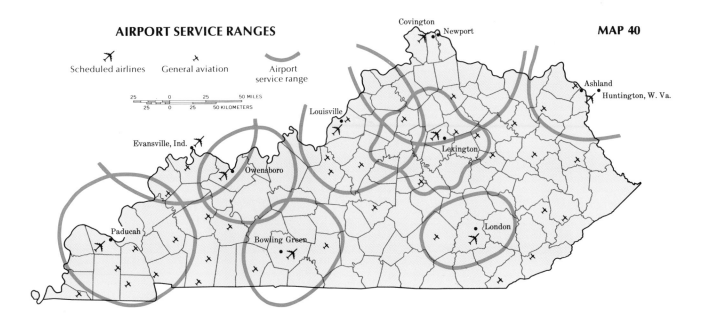

MAP 40

AIRPORT SERVICE RANGES

Scheduled airlines   General aviation   Airport service range

25   0   25   50 MILES
25   0   25   50 KILOMETERS

In eastern Kentucky water transportation is confined to the Ohio and Big Sandy rivers. The nine-foot channel on the Big Sandy extends from its mouth at the Ohio River for about nine miles upstream and the six-foot channel continues for about another eight miles. The principal commodities moved on the Big Sandy are petroleum products.

Major river ports having barge or towboat terminals are Hickman, Paducah, Henderson, Owensboro, Louisville, Covington-Newport, and Ashland. In addition to substantial commercial activity on waterways, there is considerable recreational activity, as discussed in Chapter XIII.

Various types of airport service are scattered across the state **(Map 40)**. The sixty-eight commercial airports have runways ranging from as short as 1,800 feet to as long as 9,500 feet. **Map 40** shows only the major air carrier and general aviation airports. Six large airports with commercial airline service lie near Kentucky's Ohio River border (Paducah, Owensboro, Louisville, and Covington, plus Evansville, Indiana, and Huntington, West Virginia). Lexington provides the most notable internal focus away from the Ohio River, while London in the southeast and Bowling Green in the southwest provide commercial air services for southern Kentucky communities. The service range for each airport served by commercial air carriers is indicated on the map.

Lesser but important general aviation services are available at a series of paved airfields located at intermediate and smaller Kentucky cities across the state. Twenty-four of these paved airfields are lighted, in general those serving the larger intermediate communities.

Of the large commercial airports, Louisville's Standiford Field is served by eight airlines; Covington Airport has six airlines; Lexington has four; Henderson-Evansville, three; and Ashland-Huntington, two. Paducah, London, Bowling Green, and Owensboro are served by one airline each. A recently instituted commuter airline service now provides connections between Owensboro, Louisville, Lexington, Frankfort, Covington, and Bowling Green.

Annual passenger volume, according to recent data, varies from 3,000 passengers emplaning in London, the lowest total for any of the large commercial airfields, to as high as 1,222,000 at Covington. Louisville's volume approached 900,000, while Evansville and Lexington each had well over 100,000 passengers emplaning and Bowling Green had 6,200 passengers. Completion of the new passenger terminal facility and expansion of the cargo area at Lexington's Blue Grass Field in 1976 has prompted increases in both passenger and freight traffic, has increased the air service range of Lexington, and has made Blue Grass Field more competitive with Louisville and Covington.

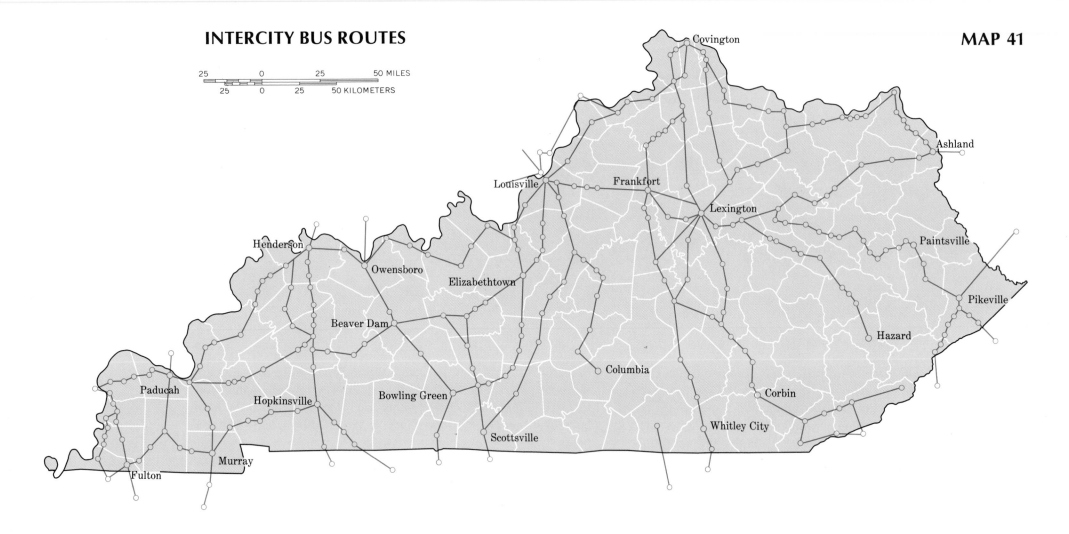

Access to scheduled intercity bus service is within twenty-five miles of even the most remote communities in Kentucky. While the number of bus-miles in scheduled service has been stable for some years, charter service is increasing 4.6 percent annually.

All Kentucky cities with populations of 5,000 and above, except some independent municipalities which are parts of metropolitan areas, are served by scheduled intercity bus service **(Map 41)**, as are most communities of 1,000 to 5,000. Access to scheduled bus service is within twenty to twenty-five miles in most parts of the state. Bus transportation is rather circuitous between some points, however. Travel from Hazard to Corbin, for example, would require going north to Lexington and then south to Corbin. The daily scheduled intercity bus-miles, estimated at 23,000, remained stable between 1970 and 1975, although the number of passengers declined. Chartered bus service has grown rapidly.

HIGHWAY TRAFFIC FLOW

MAP 42

Annual average 24-hour traffic

500–2499   2500–4999   5000–9999   10,000–14,999   15,000 and more

Traffic volume measured in average daily trips

25   0   25   50 MILES
25   0   25   50 KILOMETERS

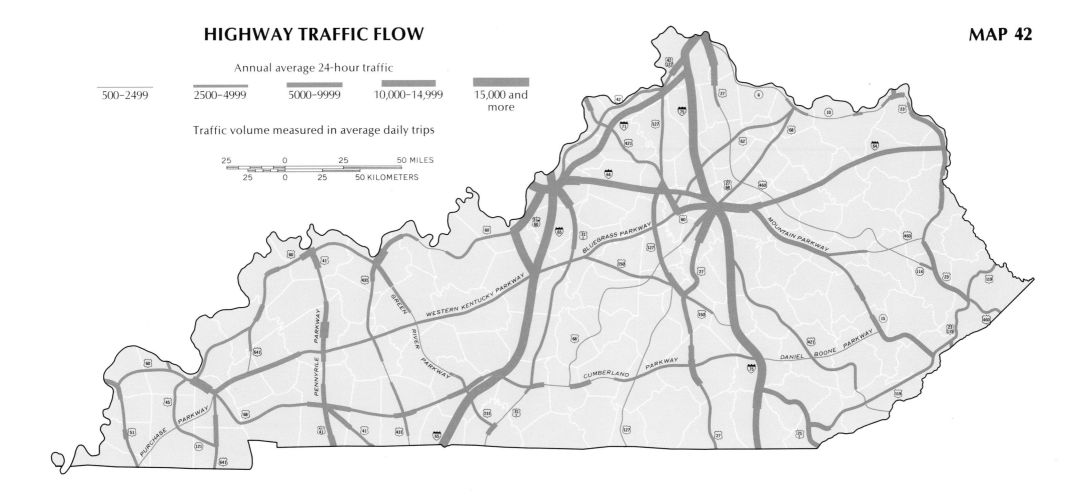

Highway travel in Kentucky has shown a dramatic increase since 1940. Greater use of the automobile for travel, the geographic dispersion of markets and economic activities, and the Interstate Highway program have all contributed to this trend. Between 1940 and 1972 Kentucky's population increased by 16 percent, while the number of registered vehicles increased by 309 percent. This represents an increase of 254 percent in the number of vehicles per person, from .168 to .595. The even more dramatic increase in annual travel per capita (431 percent) can be attributed not only to the increases in population and vehicle ownership but also to greater mobility. The increase in passenger-miles in Kentucky (434 percent) has been almost twice as much as the national average (250 percent). Freight traffic has also increased considerably.

The highways, corridors, and circles or bypasses of Kentucky form a complex road pattern (**Map 42**). The traffic flows and the routes along which they move can be viewed as clusters of major and minor routes. The heaviest flows are along some of the most highly developed routes. This is particularly true for three routes: I-75, extending south through Kentucky from Greater Cincinnati to the Tennessee border south of Williamsburg; the I-71 Cincinnati–Louisville and I-65 Louisville–Nashville routes; and I-64 from near Ashland to Louisville via Lexington's northern fringe.

The highway flows may be viewed as falling into three categories: a large number of approximately north–south routes; a considerably lesser number of essentially east–west routes; and an increasing number of locally important and heavily used bypasses or circles drawing through traffic away from the congested central areas of most large and many intermediate-size cities.

Interstate 64 (*left*) provides an east–west link from Ashland to Louisville via Lexington. Interstate routes are mainly used by automobiles, freight and mail trucks, and buses. *Right:* A highway service area near Cave City. Located near the intersection of I-65, U.S. 35W, and state routes 70 and 90, Cave City is a tourist center for Mammoth Cave National Park visitors.

At least seventeen U.S., interstate, and Kentucky parkway routes carry traffic through Kentucky in a generally north–south direction. Five east–west highways are important corridors. A northern series of Kentucky, interstate, and U.S. routes roughly parallels the Ohio River from Ashland northwest to the Covington area, southwest to Louisville, and beyond to Paducah and the Jackson Purchase. Interstate 64 from Ashland to Louisville via Lexington provides a high-speed east–west route. In recent years a series of Kentucky parkways has connected eastern Kentucky in the Prestonsburg area westward via Lexington and Elizabethtown to the Jackson Purchase at Fulton in the extreme southwest. The Cumberland and Daniel Boone parkways together provide toll roads between Bowling Green and Hazard. Finally, along the southeastern fringe of Kentucky, U.S. 119 provides a nearly continuous northeast–southwest passageway immediately north of the Virginia and Tennessee borders.

A third and increasingly vital category of highway corridors (not shown on Map 42) consists of as many as twenty circle and bypass routes diverting heavy through traffic away from the centers of many of Kentucky's cities. For other communities the interstate and Kentucky parkway sections perform the same function.

A few key highways have by far the heaviest volumes of average daily vehicles. Notable among these are the routes across the Ohio River connecting Greater Cincinnati with northern Kentucky, the routes across the Ohio River into Louisville, and New Circle Road and some of its spoke-like connecting routes serving the Lexington area. The heaviest corridors extending considerable distances across the state are I-75 from Cincinnati to the Tennessee line; I-64 from Louisville southwestward past Bowling Green to the Tennessee line toward Nashville; I-71, the transverse connector route now providing high-speed access between Louisville and northern Kentucky and the nearby Greater Cincinnati area; and

I-64 eastward from Louisville to Lexington and, to a somewhat lesser extent, onward to the Ashland-Huntington area.

The routes carrying traffic southward from the Ohio River line out of Owensboro, Henderson, Paducah, and Wickliffe are only slightly less heavily traveled. U.S. 23 in extreme eastern Kentucky from South Portsmouth and Ashland southward through the Appalachian coal field into Virginia is almost as heavily used.

In general, the east–west routes, except for limited stretches, are less heavily traveled than are the principal north–south routes. Until the recent development of the interstate and Kentucky parkway systems, good east–west connections were almost nonexistent. The principal flows of heavy-volume passenger and freight highway traffic in west-central United States pass through Kentucky along several of its main north–south routes.

WILLIAM A. WITHINGTON and WILFORD A. BLADEN

Left: The intersection of Mountain Parkway and state route 15 near Slade. *Right:* Daniel Boone Parkway in Clay County. These recently constructed modern toll roads, together with good secondary access roads, are reducing the time-distance factor and helping to break down the traditional physical isolation of Appalachian Kentucky. Many persons who were raised in this region but work in industrial cities in Ohio, Indiana, and Michigan are now able to commute home easily on weekends. Plans for the development of Appalachia have emphasized the building of modern highways to generate economic activity and attract new industries to the area.

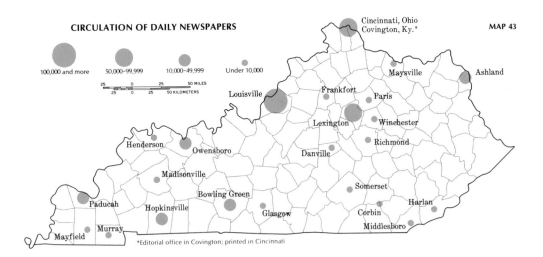

MAP 43

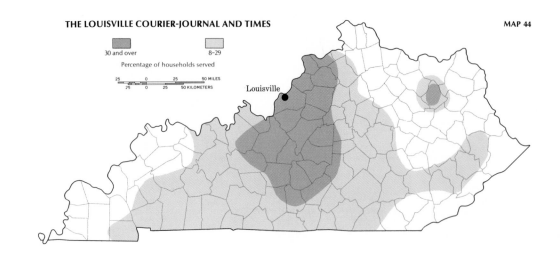

MAP 44

MAPS 43-46

NEWSPAPER CIRCULATION

The mass media have been compared to the nerve system of the human body in the way they spread information through our society. Newspapers, television, and radio influence the lives of all Americans. The access to mass media varies greatly from place to place in Kentucky. Major metropolitan centers have traditionally had a wide range of newspapers and television and radio stations available, while some rural areas have lacked these amenities. Because each newspaper or broadcasting station originates in a particular place, it generates an "information field" about that place. A reader of the Louisville *Courier-Journal*, for example, will learn a great deal about Louisville from the news and advertising, while a reader of the Lexington *Herald-Leader* will find out much more about Lexington. Because of the information fields generated, mass media communications affect the retail trade hinterlands of cities.

Daily newspapers are published in twenty-two Kentucky cities;* their circulation is shown on **Map 43.** These newspapers vary wide-

* The *Kentucky Post*, which has a circulation of about 59,000 in northern Kentucky, has its editorial office in Covington but is printed in Cincinnati. Its circulation area is shown on Map 46.

ly in circulation and thus in the number of persons served. Fifteen cities have newspapers with circulations of less than 10,000; five have newspapers with total circulations of 10,000 to 50,000. The total circulation of Lexington newspapers is about 90,000, while that of Louisville newspapers is about 410,000.

Maps 44–46 show the in-state circulation territories of three types of newspapers: the Louisville *Courier-Journal* and *Times*, other in-state papers with over 10,000 circulation each, and out-of-state newspapers. In each map two areas are shown: those in which a newspaper reaches over 30 percent of all households and those in which it reaches between 8 and 29 percent. Where a newspaper reaches over 30 percent of the households it is usually dominant over all other newspapers, while 8 percent of the households is a general cut-off point for same-day carrier delivery as opposed to mail delivery.

The Louisville *Courier-Journal* and *Times* **(Map 44)** have by far the largest circulations of any newspapers in the state. The morning *Courier-Journal* is unique among state newspapers in that it circulates in almost all of Kentucky's counties. The areas dominated by the *Courier-Journal* and *Times* are generally those within fifty

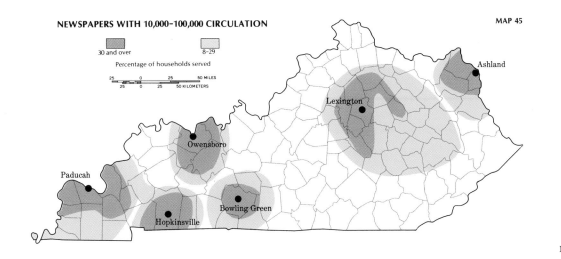

30 and over   8-29
Percentage of households served
25   0   25   50 MILES
25   0   25   50 KILOMETERS

Ashland
Lexington
Owensboro
Paducah
Bowling Green
Hopkinsville

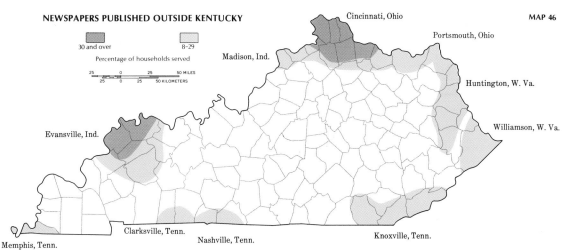

30 and over   8-29
Percentage of households served
25   0   25   50 MILES
25   0   25   50 KILOMETERS

Cincinnati, Ohio
Portsmouth, Ohio
Madison, Ind.
Huntington, W. Va.
Evansville, Ind.
Williamson, W. Va.
Clarksville, Tenn.
Memphis, Tenn.
Nashville, Tenn.
Knoxville, Tenn.

to eighty miles of Louisville. The *Courier-Journal* also dominates Rowan County, undoubtedly due to the presence of Morehead State University. Significant levels of *Courier-Journal* circulation are found in most parts of the state except the western and northeastern areas, which are quite distant from Louisville and are dominated by newspapers from nearby cities, such as Cincinnati, Ashland, or Huntington.

Each of the six centers other than Louisville that has newspapers with over 10,000 circulation dominates its particular section of the state **(Map 45)**. Ten cities outside Kentucky publish newspapers which circulate to at least 8 percent of the households in one or more Kentucky counties **(Map 46)**. Out-of-state newspapers generally circulate only in a fringe of counties around the margin of the state.

In addition to daily newspapers, Kentucky is served by many weekly newspapers, at least one in each of the state's 120 counties. The weekly newspapers are generally not a significant source of national or state news but keep people well informed of local events.

PHILLIP D. PHILLIPS and WILLIAM A. WITHINGTON

MAP 47

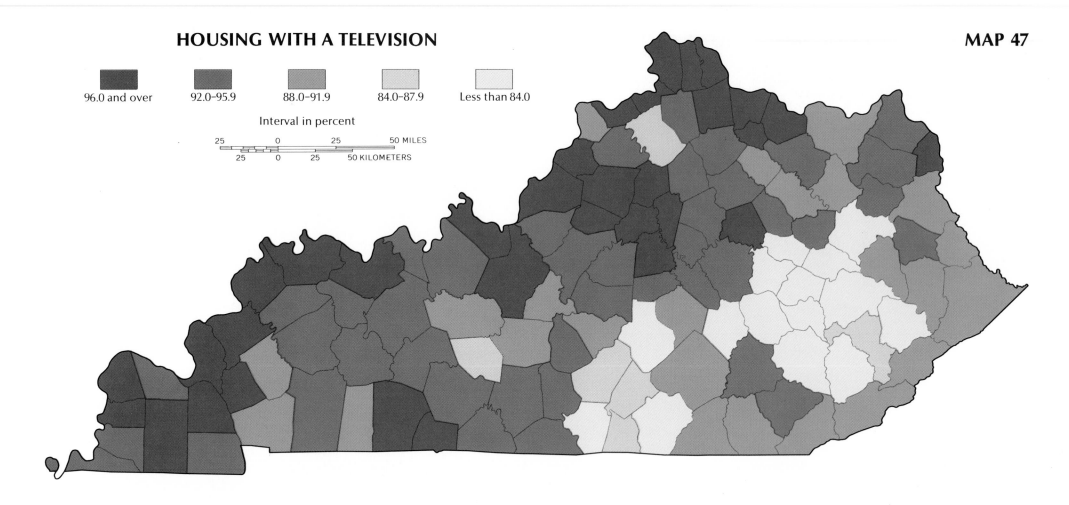

HOUSING WITH A TELEVISION

96.0 and over   92.0–95.9   88.0–91.9   84.0–87.9   Less than 84.0

Interval in percent

25   0   25   50 MILES
25   0   25   50 KILOMETERS

MAP 47

TELEVISION

The increasing availability of television programs from one or more stations to most households in Kentucky is reflected in the fact that 94 percent of Kentucky's nearly 984,000 occupied housing units had one or more television sets in 1970. **Map 47** indicates their distribution by county. Indirectly, this map suggests the intensity and diversity of television coverage and the amount of television viewing in the state.

Proportions of television sets vary somewhat between urban and rural areas and between various regions of Kentucky. In urban areas, which contain more than half the total population, 95.9 percent of all homes had at least one television set in 1970. In rural areas, the level was slightly lower, 91.7 percent. The statewide

patterns by counties showed the lowest proportions in Breathitt County. The highest levels were concentrated largely in counties immediately south of the Ohio River across the entire northern fringe of the state, from the Jackson Purchase eastward to Boyd County. Only Trimble, Owen, and Lewis counties in northern and northeastern Kentucky were not in the two highest levels.

In contrast, Kentucky had several areas with relatively low proportions of television sets. The two largest areas were in southeastern and south-central Kentucky. Edmonson County north of Bowling Green and Owen County halfway between Louisville and Covington were two individual and widely separated counties with below-average percentages of television sets. These areas are

characterized primarily by rural or small-town populations and are predominantly in hilly to mountainous areas away from cities of 10,000 or more. In no Kentucky county with a city of 5,000 or more did less than 88 percent of the occupied household units have television sets. The number of household units with one or more television sets has increased since 1970.

Television reception in Kentucky comes from thirty-two commercial stations, eleven of which are in the state. In addition, the Kentucky Educational Television (KET) network, a public system, provides statewide coverage. Of the eleven commercial stations within Kentucky, five broadcast from Louisville, three from Lexington, and one each from Paducah, Bowling Green, and Hazard.

WILLIAM A. WITHINGTON and WILFORD A. BLADEN

Kentucky Educational Television headquarters (*top right*) is located in Lexington. The network has thirteen transmitters broadcasting cultural and educational programs. The science series "The Universe and I" (*left*) is telecast to junior high school students throughout the state. WKYT-TV (*above*) is one of eleven commercial stations in Kentucky. Operating with a power of 2,360,000 watts, WKYT transmission covers the Bluegrass Region of central Kentucky. Interest in television was evident in Kentucky as early as 1929 when WHAS, Louisville, began experimentation with a power of only 10,000 watts.

Georgetown (*above left*), incorporated in 1790 and named for George Washington, entered the urban hierarchy by 1800. Today it is a thriving Bluegrass city, the home of Georgetown College, established in 1829, and has recently attracted several industries. Maysville (*above*), located on the Ohio River, was established in 1787 and by 1792 was the leading port of entry for settlers coming to Kentucky. It entered the urban hierarchy by 1840. Maysville's early prosperity was based on river traffic. Merchandise came down the Ohio on barges to Maysville and was hauled overland to the growing cluster of settlements in central Kentucky. Elkhorn City (*left*) is a coal-mining town near the Kentucky-Virginia line in Pike County. It entered the urban hierarchy after 1910 but fell below the urban threshold by 1950.

VIII. THE URBAN SYSTEM

EVOLUTION OF THE URBAN HIERARCHY

In 1970 for the first time Kentucky had a population more urban than rural, according to the United States Census Bureau. From the earliest days of white settlement in the eighteenth century, however, Kentucky has had an evolving urban pattern. Although specific urban or settlement population figures for 1790 were not generally recorded for the Kentucky District of Virginia, Lexington had a population of 835. A number of other Inner Bluegrass settlements also had small urbanlike populations. **(Maps 48–53** use present county boundaries.)

Questions as to the size of population, the mixture of activities, or the kinds of structures or facilities needed if a settlement is to be identified as urban are important and difficult to answer. In the present context, settlements are counted as urban which have a population for each census period which is above an urban threshold (T_y) determined with minor adjustments by the equation: $T_y = T_o (P_y / P_o)$. The Census Bureau's urban threshold (in use since 1910) of 2,500 (T_o) is multiplied by Kentucky's population in the particular census year (P_y) divided by the Kentucky population of 1970 (P_o). **Table 6** gives the urban threshold for each of seven census decades from 1800 to 1970 and the number of places surpassing these thresholds.

Using the above determination for urban places, Kentucky in 1800 had twelve settlements with at least 170 people each. By 1870 the total population of the state had increased sixfold and the urban threshold exceeded 1,000. By 1930, Kentucky's urban threshold had doubled to just over 2,000 people; by 1970, the national urban threshold of 2,500 was also that of Kentucky. The number of

places identified as urban by this threshold formula nearly doubled between 1800 and 1840; by 1870 the number increased by more than 50 percent; and by 1910 it increased again by nearly 50 percent. Since 1910, the increases in the number of urban places have been somewhat less rapid: about 29 percent between 1910 and 1930; only about 19 percent between 1930 and 1950; and just over 27 percent from 1950 to 1970. Rapid increases in suburban communities near major urban centers were notable during this last twenty-year period.

In 1800 **(Map 48)** the twelve urban communities were most con-

TABLE 6
Kentucky Urban Population Threshold, 1800-1970

Census year	Population threshold	Number of urban places[a]
1800	170	12
1840	605[b]	21
1870	1,022[c]	36 (34)
1910	1,777	52 (43)
1930	2,030	67 (58)
1950	2,285	80 (66)
1970	2,500	102 (78)

[a]Numbers in parentheses are numbers of places shown on Maps 48-54. These places in some instances represent groupings within a local area.

[b]If the 1840 minimum had been reduced to 600, the two Ohio River ports of Port William and Warsaw would have been included, but neither place ever attained threshold size in succeeding censuses.

[c]For 1870 the equation figure of 1,025 was changed to 1,022 to include Madisonville in west-central Kentucky, which continued as an increasingly sizable and important urban center.

centrated in the Inner Bluegrass and along the route to Louisville. Louisville and three smaller ports, characterized by Ohio River border sites, accounted for a third of all urban places. By 1840 (**Map 49**), more than half of the twenty-one urban places were in the Inner Bluegrass. Others lay along the Ohio River and between the Inner Bluegrass and Louisville. Four other new urban communities represented the beginning of urbanization in west-central Kentucky away from the Ohio River artery.

By 1870 urbanization had spread into all but the southeastern and south-central parts of Kentucky (**Map 50**). The Bluegrass centers still dominated. Many more cities, however, from Catlettsburg in the east where the Big Sandy joins the Ohio, to Columbus and Hickman in the far west on the Mississippi, were sited along Kentucky's principal fringing navigable rivers. Although still limited in number, at least five Kentucky cities were in west-central areas away from the Ohio-Mississippi waterways. Russellville and Hopkinsville had surpassed the urban threshold by 1840. By 1870, Madisonville, Bowling Green, and Franklin in this west-central area also rose above the urban threshold.

The forty years between 1870 and 1910 witnessed the spread and intensification of urbanization in Kentucky to include fifty-two cities (**Map 51**). The nine cities added in west-central and far western Kentucky made up the single largest group of new urban places. Five cities appeared as the first for southeastern Kentucky. The remaining new urban places represented further intensification of urban development in northern Kentucky and in Jefferson County near Louisville.

The most notable change in urbanization between 1910 and 1930 (**Map 52**) was the appearance of several southeastern Kentucky cities in the one major area with almost no urban places previously. The few other new urban centers were satellites of Covington-Newport in the northern Kentucky cluster of cities and in west-central Kentucky.

In contrast to 1930, Kentucky's 1950 pattern of urbanization (**Map 53**) reached into most areas of the state. In eastern Kentucky more cities reached the urban threshold, but others previously urban had lost that status. The Bluegrass and its fringes had the largest number of additions between 1930 and 1950. Two Lexington suburban subdivisions, later part of the city of Lexington, as well as Berea, Lancaster, and Wilmore to the south, Morehead to the east, and Lawrenceburg to the west were part of this pattern of urban growth. Outlier communities in northern Kentucky and the Louisville area continued the suburbanization trend around

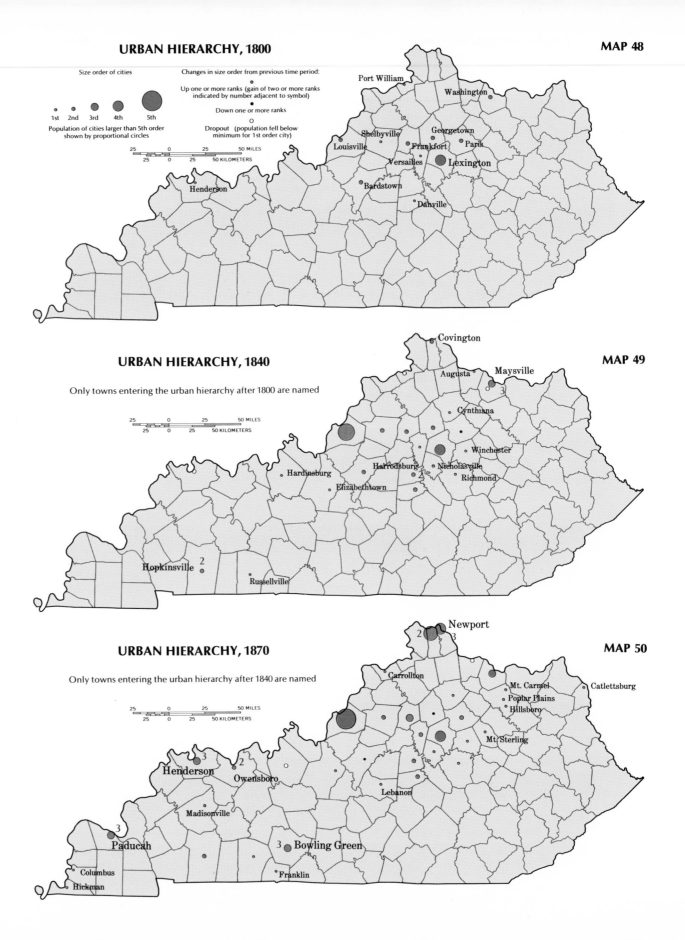

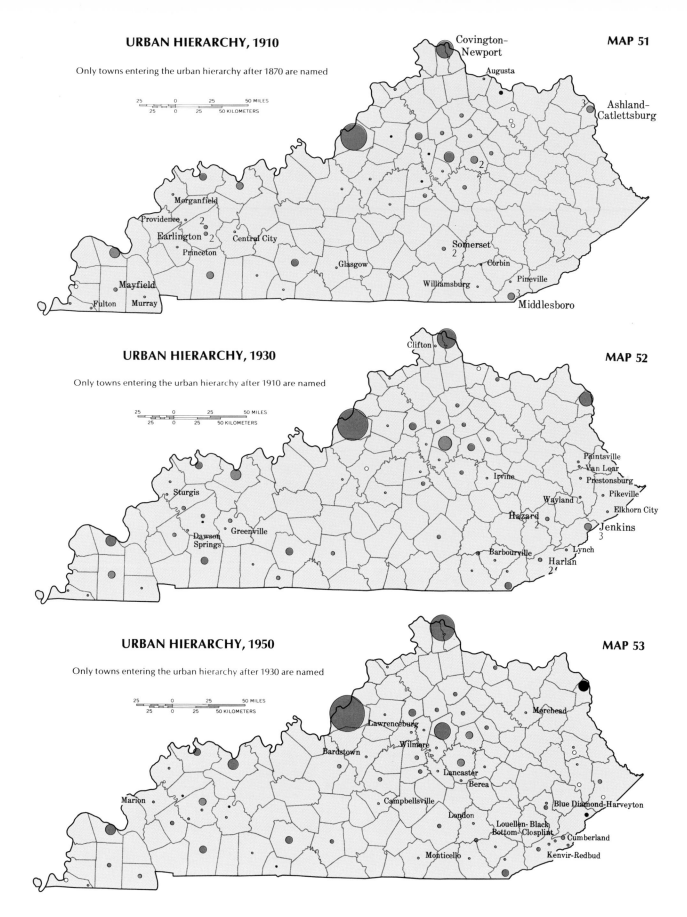

URBAN HIERARCHY, 1910

Only towns entering the urban hierarchy after 1870 are named

25 0 25 50 MILES
25 0 25 50 KILOMETERS

MAP 51

Covington-Newport
Augusta
Ashland-Catlettsburg
Morganfield
Providence
2
Earlington
2
Central City
Princeton
Somerset
2
Glasgow
Corbin
Williamsburg
Pineville
Mayfield
Murray
Fulton
Middlesboro
3

URBAN HIERARCHY, 1930

Only towns entering the urban hierarchy after 1910 are named

25 0 25 50 MILES
25 0 25 50 KILOMETERS

MAP 52

Clifton
Paintsville
Van Lear
Prestonsburg
Sturgis
Irvine
Pikeville
Wayland
Elkhorn City
Hazard
2
Jenkins
3
Greenville
Dawson Springs
Barbourville
Lynch
Harlan
2

URBAN HIERARCHY, 1950

Only towns entering the urban hierarchy after 1930 are named

25 0 25 50 MILES
25 0 25 50 KILOMETERS

MAP 53

Morehead
Lawrenceburg
Wilmore
Bardstown
Lancaster
Berea
Marion
Campbellsville
Blue Diamond-Harveyton
London
Louellen-Black Bottom-Closplint
Monticello
Cumberland
Kenvir-Redbud

Mayfield, seat of Graves County, is the most important of the smaller cities in the Jackson Purchase. It is a major market for farm products, including dark fire-cured tobacco, and has several men's clothing factories. Mayfield was founded in 1823 and entered the urban hierarchy after 1870. It had registered several gains in urban rank by 1970.

larger and older urban centers. The reemergence of Bardstown and achievement of urban status for Monticello to the south filled in south-central Kentucky urbanization. Marion was the only new urban place west of the Louisville-Jefferson County area. Since two western Kentucky places near the Ohio and Mississippi rivers lost urban status, there was no net change in urbanization for this area.

In the most recent census period, from 1950 to 1970 (**Map 54**), urbanization in Kentucky has had several facets. Four coal-mining communities in the southeast lost urban status. Both Fort Knox and Fort Campbell by census definition became urban places. The spread of urbanization around existing major urban centers accounted for the largest part of other urban growth in the northern Kentucky-Greater Cincinnati area, in the Louisville-Jefferson County area and to its south, and near Ashland and Paducah. The devel-

MAP 54

URBAN HIERARCHY, 1970

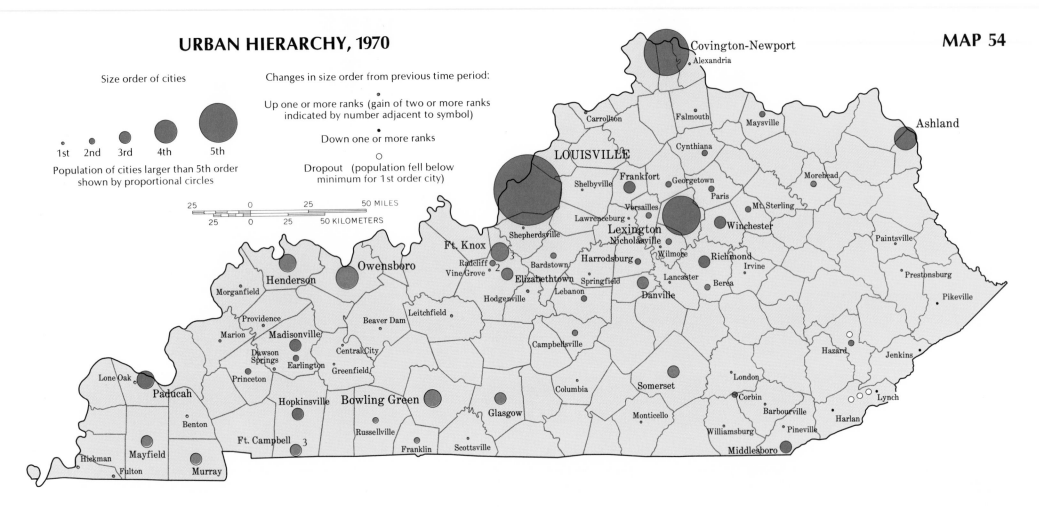

Size order of cities

1st 2nd 3rd 4th 5th

Population of cities larger than 5th order
shown by proportional circles

Changes in size order from previous time period:

Up one or more ranks (gain of two or more ranks
indicated by number adjacent to symbol)

Down one or more ranks

Dropout (population fell below
minimum for 1st order city)

25   0   25   50 MILES
25   0   25   50 KILOMETERS

opment of high-speed parkways may account, at least in part, for a string of urban center additions which are along or near the line of the Purchase, Western, and Blue Grass parkways. As of 1970 these 102 urban centers of 2,500 population or larger were in sixty-nine of Kentucky's 120 counties, leaving fifty-one counties, 42.5 percent of the total number, still without an urban place. Only fourteen counties, however, less than 12 percent of the total, were without an "urban-like" place of 1,000 or more population, the level listed in Census Bureau publications since 1910. Of these counties, five are scattered across western and central Kentucky, and nine are in eastern Kentucky.

By 1970, Kentucky's urbanization included six metropolitan areas: Louisville-Jefferson County and adjacent Indiana counties; Boone, Kenton, and Campbell counties south of Cincinnati; Ashland-Boyd County in the Ashland-Huntington area; Owensboro-Daviess

County; Henderson (city and County) in the Evansville, Indiana-Kentucky area; and Lexington-Fayette County. Metropolitan urbanization has expanded since mid-1973. Lexington and Fayette County are most affected, with five of the six peripheral counties being added. Two Kentucky counties north and south of Jefferson County expand the greater Louisville metropolitan area; Greenup County was added to Boyd County in the Ashland area, and Hopkinsville-Christian County in the southwest became metropolitan. (Table 3, p. 18, gives more information on these areas.)

In summary, Kentucky today is more urban than rural. According to current census definitions, however, large areas and populations remain nonurban. Notably limited in urban development is eastern Kentucky.

WILLIAM A. WITHINGTON and PHILLIP D. PHILLIPS

MAP 55

KENTUCKY COMMUTER SHEDS

Total number of commuters by county:

under 100   100-249   250-499   500-999   1000-2499   2500-4999   5000 and over

Central county of metropolitan area

25   0   25   50 MILES
25   0   25   50 KILOMETERS

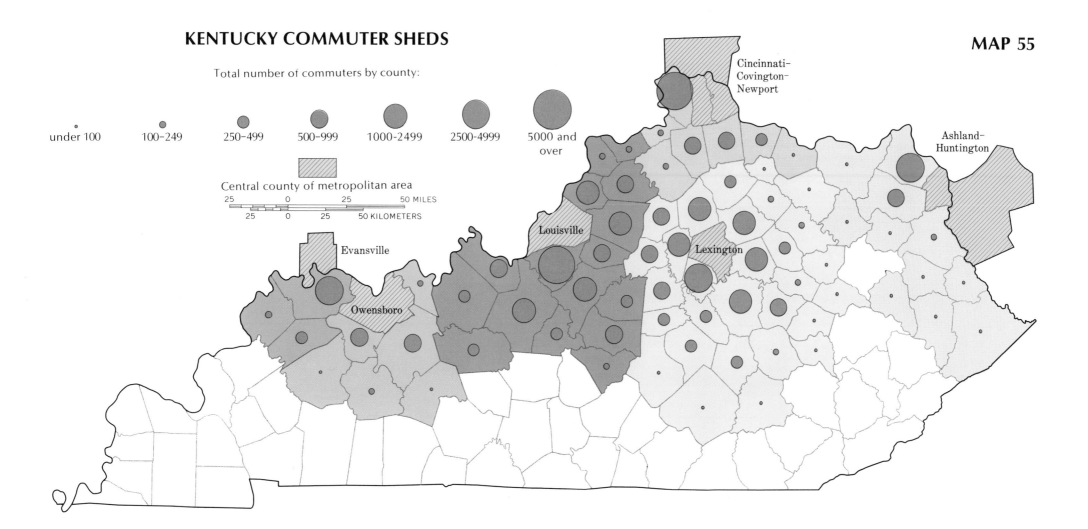

Cincinnati-Covington-Newport

Ashland-Huntington

Evansville

Louisville

Lexington

Owensboro

COMMUTER SHEDS

MAPS 55-58

Workers around Kentucky's cities move in a "daily tidal flow" to and from surrounding small towns and rural areas. Six metropolitan centers received about 80,000 daily commuters in 1970 (**Map 55**). In Bullitt and Boone counties over 5,000 workers commute to nearby metropolitan centers, while persons in seventy-eight of Kentucky's counties commuted to at least one of these six centers, according to the 1970 census. In some counties there was commuting to two or three centers. In Owen County, for instance, persons commuted to Louisville, Lexington, and Cincinnati-Covington-Newport.

Many persons commuting to metropolitan centers live in small towns while others live in rural areas. Some are natives of the areas in which they live who have stayed there because of ties with the area and a preference for small-town or rural life. Others have moved from metropolitan centers, also seeking the perceived virtues of the country—clean air, open space, and safety, among others. These latter persons may be referred to as "exurbanites." Persons in both groups commute from 40 to 200 miles round-trip each day. Because of this, they are particularly vulnerable to increases in gasoline prices and would find their lifestyle jeopardized by gasoline rationing or price increases. Commuters are a significant population element in many nonmetropolitan counties and if they were forced to move closer to their jobs it would severely affect both metropolitan and nonmetropolitan areas. The non-

Over 90 percent of Kentuckians live within an hour's drive of a metropolitan area where they shop and work. The result is a large volume of commuter traffic, such as that (*above*) on Interstate 75, which carries an estimated 78,000 persons daily. *Above right:* Local commuter traffic in the south end of Lexington. Nearly 15,000 persons commute daily to Lexington from nearby counties and over 2,000 commute out of the city, primarily to Franklin County.

metropolitan areas would be faced with depopulation, loss of income, and loss of business for local merchants, while metropolitan areas would be hard pressed to provide housing, schools, and other facilities for the newcomers. Commuters, their families, and the "secondary employment" of merchants, teachers, and other service people that they support represent about 400,000 persons, or one-eighth of the population of Kentucky. The areas of dominant flow into these centers might be thought of as Kentucky's "daily urban systems."

The proportions of workers residing in various counties and commuting to the six major urban areas are shown in **Maps 56–58**. The reach of many of these centers in drawing commuters—over 100 miles in some cases—is apparent. The commuter shed of Lexington **(Map 57)** illustrates the importance of good roads and employment opportunities in the extent of commuting that occurs. Interstate highways and parkways help to bring commuters to Lexington from such distant places as Bullitt and Pulaski counties, while in some Mountain counties with very limited local employment opportunities, over 10 percent of the work force commute to Lexington.

PHILLIP D. PHILLIPS

METROPOLITAN COMMUTER SHEDS

Commuters to metropolitan county

 20.0 and over   10.0–19.9    5.0–9.9   1.0–4.9   0.1–0.9

Interval in percent of work force

Central county of metropolitan area

50   0   50   100 MILES
50   0   50   100 KILOMETERS

OWENSBORO
(DAVIESS CO.)
LEXINGTON
(FAYETTE CO.)

MAP 57

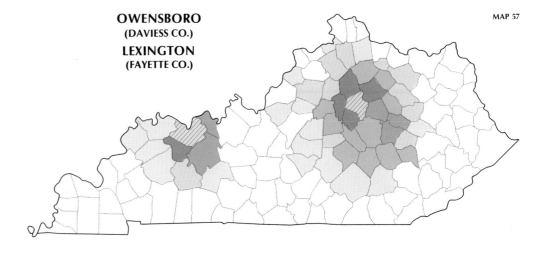

LOUISVILLE
(JEFFERSON CO.)
ASHLAND-HUNTINGTON
(BOYD CO., KY., AND CABELL AND
WAYNE COUNTIES, W. VA.)

MAP 56

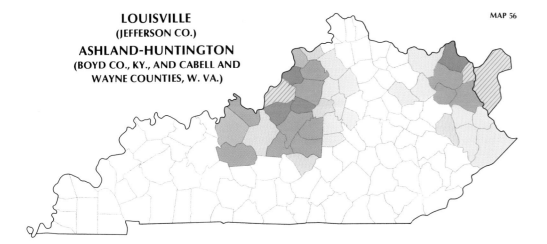

CINCINNATI-COVINGTON-NEWPORT
(HAMILTON CO., O., AND KENTON
AND CAMPBELL COUNTIES, KY.)
EVANSVILLE
(VANDERBURGH CO., IND.)

MAP 58

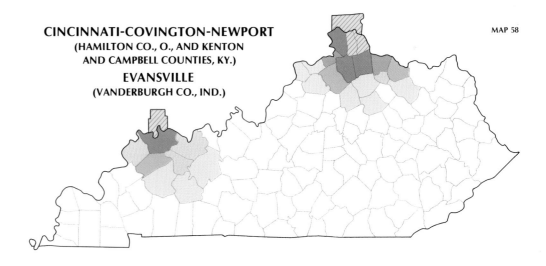

Kentucky's largest cities display a variety of landscapes and urban problems. *Above left:* The heavily urbanized landscape of Louisville. Once a blighted area, the inner section of the city has undergone extensive urban renewal. *Left:* Covington and the bridge leading to downtown Cincinnati. Older housing areas along the Ohio River have been replaced by motels and other highway service establishments near Interstate 75. Ashland's older housing areas are located in the central portion of the city (*above*), close to the Ohio River. Residential areas with newer housing have developed on the bluffs away from the river. *Opposite:* A black residential area in Lexington's inner city. The number of blacks in inner cities increased in the 1960s while the white population declined. The heaviest shifts occurred in Louisville and Lexington.

STRUCTURE OF MAJOR URBAN AREAS

Physical and social conditions and historical development divide any city into a myriad of small areas. The overall perspective provided by **Maps 59–61** allows one to generalize about the urban structure. Not only do the maps make the structure more evident, they also make the processes that create it more interpretable.

Three aspects of the spatial structure of Kentucky's major cities have been chosen for presentation here: age of housing, income, and race. These variables represent the three most important dimensions of urban structure. The age of housing reveals the sequence of development of the cities and also indicates a great deal about the style and condition of the housing. Age of housing is also closely associated with the age structure of the population, with older persons generally being found in areas of older housing. Income, the second variable, is probably the most important single measure of social status in American society today. The amount of money that families have available to them is an important indicator of many aspects of their lifestyle and the quality of housing, clothing, food, and other necessities and luxuries available to them. Race is an important and inescapable dimension of American urban structure because of the segregation of blacks found in all American cities. Racial differences cut across age and income differences, despite the tendency of black populations to be confined to the older and poorer sections of cities.

These maps of four major metropolitan areas of Kentucky are based on data from the 1970 census, with information shown by census tracts. The major spatial patterns portrayed here have remained substantially unchanged since 1970.

Map 59, Age of Housing, indicates differences in the general rates of growth of the four metropolitan areas. Ashland, which is growing most slowly, has the oldest general housing stock, while Lexington and Louisville have a much greater percentage of newer housing. The four cities show similar concentric patterns of housing age around the oldest nucleuses of development. The oldest housing stock is found in the central areas, the youngest in the growing suburban fringes. Beyond the suburban fringes are older housing areas of the primarily rural peripheries.

Each of the four metropolitan areas displays a distinctive pattern within the general concentric framework, resulting from differences in topographic control on housing construction and from the location of the earliest nucleus, or nucleuses, of urban growth. The oldest housing in the Ashland area is found in the central portion of Ashland and in Catlettsburg. While no areas with the highest percentage of new housing (75 percent or more built in 1950–1970) are shown, the newer areas tend to be on the bluff tops away from the Ohio River. Louisville displays the most clearly concentric pattern. The several original settlement nucleuses, principally

MAP 59

AGE OF HOUSING

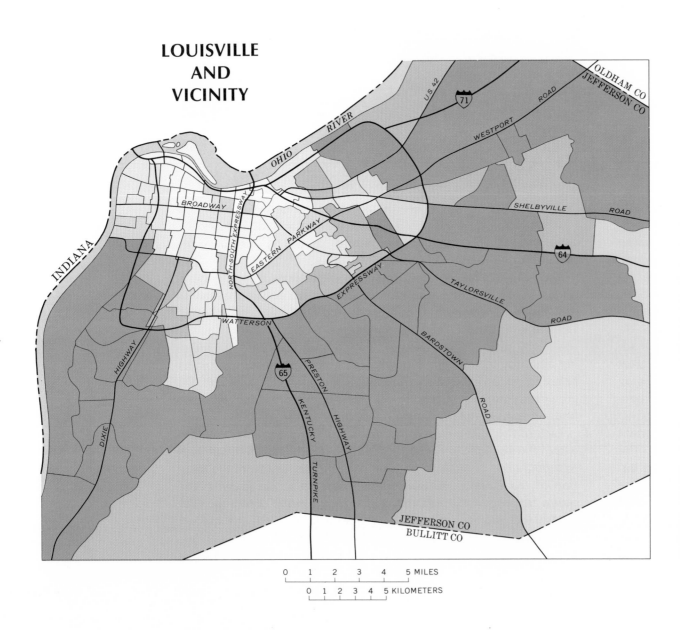

**LOUISVILLE
AND
VICINITY**

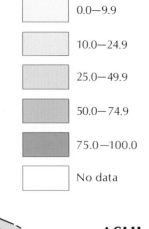

Percentage of dwelling units
constructed between 1950
and 1970.

	0.0—9.9
	10.0—24.9
	25.0—49.9
	50.0—74.9
	75.0—100.0
	No data

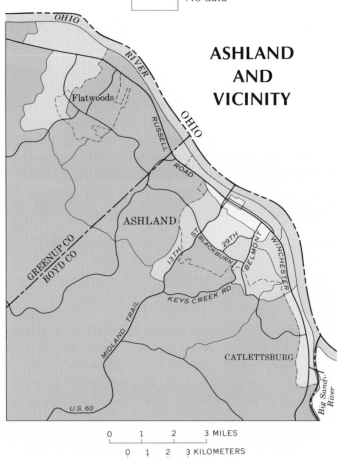

**ASHLAND
AND
VICINITY**

0 1 2 3 4 5 MILES

0 1 2 3 4 5 KILOMETERS

0 1 2 3 MILES

0 1 2 3 KILOMETERS

Louisville and Portland, have long since coalesced into a single center characterized generally by older housing. In a few central areas the housing is somewhat newer, largely as a result of urban renewal and the construction of public housing. The newest housing is found generally in a circumferential belt around the city just beyond the Watterson Expressway. Fringe areas of Jefferson County have somewhat older housing, largely rural housing not yet displaced by subdivisions.

Lexington's pattern is similar to that of Louisville. The more rapid growth of the south side than of the north side is apparent, how-

ever. Covington presents the most complex picture because of the nature of the topography and numerous early centers of urban growth. In addition to Covington and Newport, the towns of Ludlow, Bellevue, and Dayton were early urban nucleuses along the Ohio River. Elsmere also stands out as an early farming and commercial center well away from the river. In general, the highest percentage of older housing is found in the lowlands along the Ohio and Licking rivers. Above the river bluffs the housing is much newer, with the newest housing areas being in Boone County to the southwest.

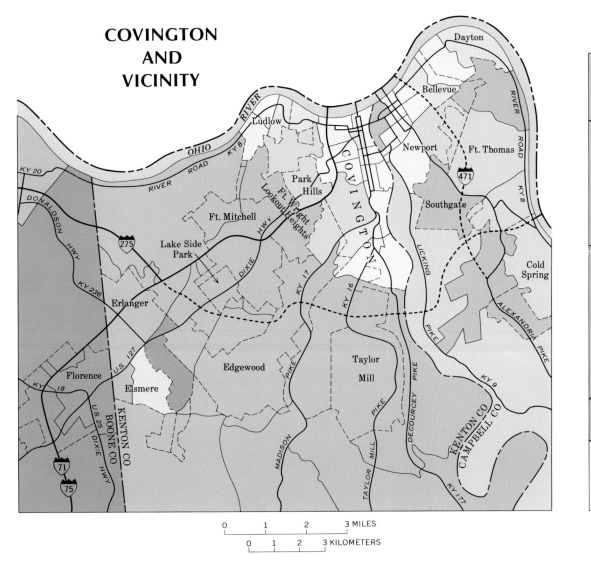

COVINGTON AND VICINITY

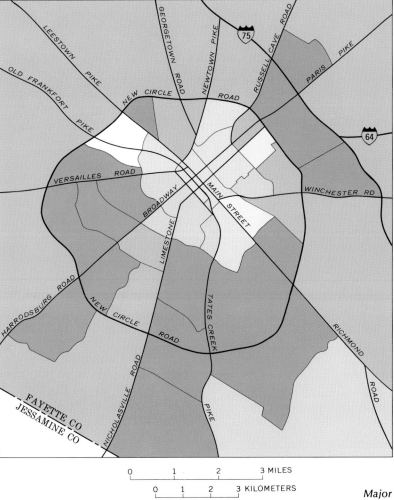

LEXINGTON

MAP 60

INCOME

Median family income, 1969

$12,000 and over

10,000-11,999

8000-9999

6000-7999

5999 and under

No data

LOUISVILLE
AND
VICINITY

ASHLAND
AND
VICINITY

OHIO RIVER

U.S. 42

71

OLDHAM CO
JEFFERSON CO

WESTPORT ROAD

SHELBYVILLE ROAD

64

INDIANA

BROADWAY

NORTH-SOUTH EXPRESSWAY

EASTERN PARKWAY

WATTERSON

EXPRESSWAY

TAYLORSVILLE

ROAD

BARDSTOWN

ROAD

65

PRESTON HIGHWAY

KENTUCKY TURNPIKE

DIXIE HIGHWAY

JEFFERSON CO
BULLITT CO

OHIO RIVER

RUSSELL ROAD

OHIO

Flatwoods

ASHLAND

GREENUP CO
BOYD CO

13TH ST

BLACKBURN

29TH

BELMONT

WINCHESTER

KEYS CREEK RD

MIDLAND TRAIL

CATLETTSBURG

U.S. 60

Big Sandy River

0 1 2 3 4 5 MILES

0 1 2 3 4 5 KILOMETERS

0 1 2 3 MILES

0 1 2 3 KILOMETERS

Income, like age of housing, varies both among and within the four major metropolitan areas **(Map 60)**. Generally, metropolitan-wide income levels tend to be high in Louisville and Lexington, intermediate in Covington, and lower in Ashland. The differences in income level reflect differences in regional location, types of business and industry in the area, and recent economic growth rates. Within the metropolitan areas, the high-income areas are generally found radiating outward from the central business district. The lowest income areas are generally found near the central business district and extending outward from it in one or more directions.

The highest-income sections of Louisville are found on the east side of town. In Lexington they are on the south side, and in Covington, to the southwest and east. In Ashland, the higher income areas are to the south and northwest. In all four metropolitan areas the low-income areas are clustered around the central business district. In Ashland and Covington topographic effects are important, with low income being found in the river valleys and high income on the bluff tops.

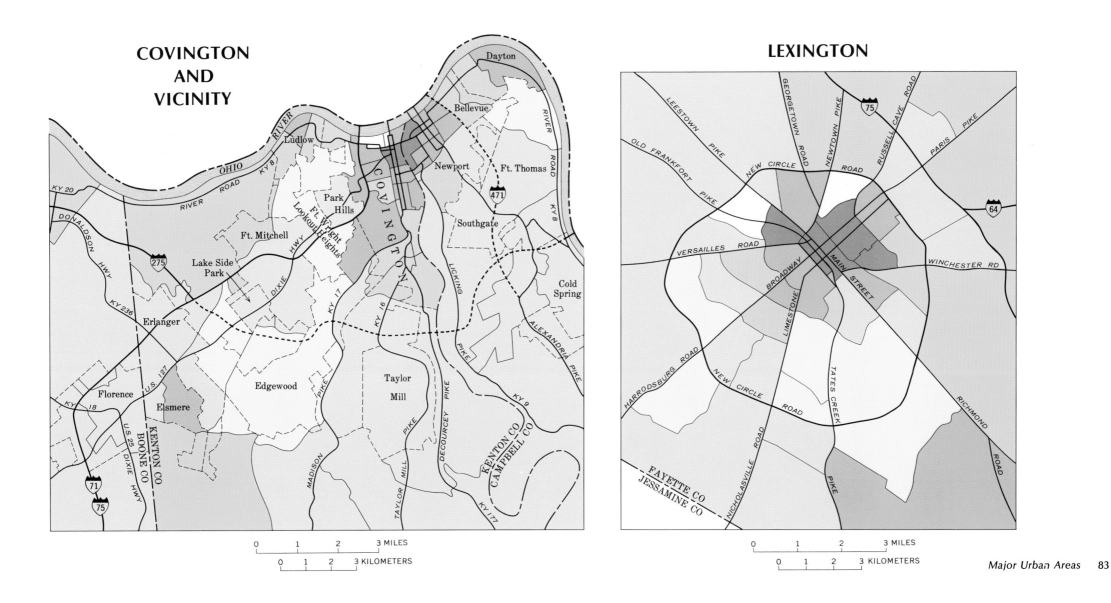

MAP 61

BLACK POPULATION

Percent of population black, 1970

	0.0—4.9
	5.0—24.9
	25.0—49.9
	50.0—74.9
	75.0—100.0

LOUISVILLE AND VICINITY

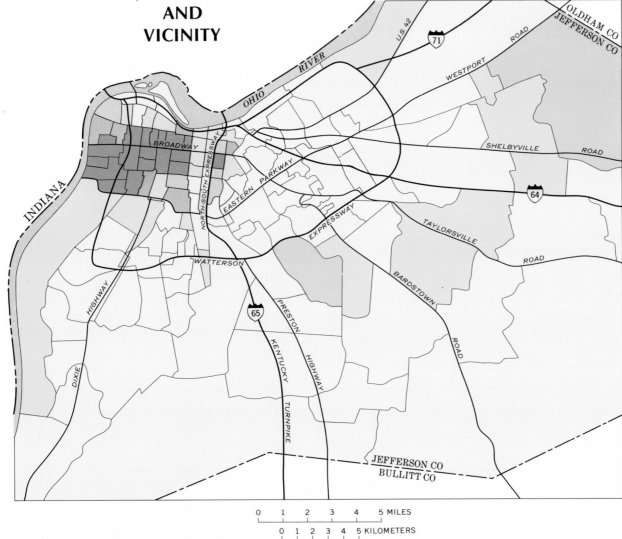

OHIO RIVER

INDIANA

U.S. 42

71

WESTPORT ROAD

OLDHAM CO
JEFFERSON CO

SHELBYVILLE ROAD

BROADWAY

NORTH-SOUTH EXPRESSWAY

EASTERN PARKWAY

64

TAYLORSVILLE ROAD

WATTERSON EXPRESSWAY

BARDSTOWN ROAD

65

KENTUCKY TURNPIKE

PRESTON HIGHWAY

DIXIE HIGHWAY

JEFFERSON CO
BULLITT CO

0 1 2 3 4 5 MILES

0 1 2 3 4 5 KILOMETERS

ASHLAND AND VICINITY

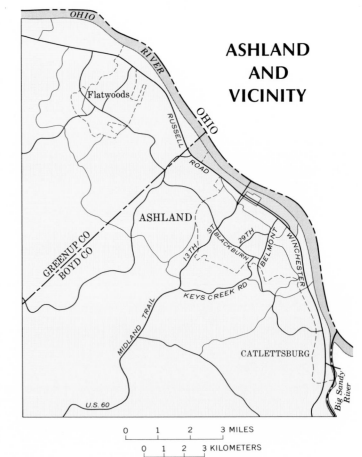

OHIO RIVER

OHIO

Flatwoods

RUSSELL ROAD

ASHLAND

GREENUP CO
BOYD CO

13TH
ST
BLACKBURN
29TH
BELMONT
WINCHESTER

KEYS CREEK RD

MIDLAND TRAIL

CATLETTSBURG

Big Sandy River

U.S. 60

0 1 2 3 MILES

0 1 2 3 KILOMETERS

84

The black population of Kentucky's major metropolitan centers is highly concentrated within certain portions of the cities (**Map 61**), as is the case in all American metropolitan areas. Also, the proportion of the population that is black varies considerably, from 1.6 percent in Ashland to 13.8 percent in Louisville. The highest proportions of black population, 75 percent or more, are generally found in older, central city areas, with more than 50 percent of the total in Louisville and Lexington.

The core of the Louisville black ghetto is in the western portion of the central city, while the core of the Lexington ghetto includes several fragmented areas generally to the north and east of the central business district. While the proportion of the population that is black is much smaller in the Ashland and Covington-Newport areas as a whole, the concentration of blacks in central city areas is apparent here, as well. Along the fringes of Lexington and Louisville, black populations range from 5.0 to 24.9 percent. In the case of Lexington, the city has spread onto the older rural black hamlets. In the Louisville area, some older black hamlets on the fringes of the city have begun to grow as a result of the influx of suburbanizing blacks.

PHILLIP D. PHILLIPS and WILLIAM A. WITHINGTON

COVINGTON AND VICINITY

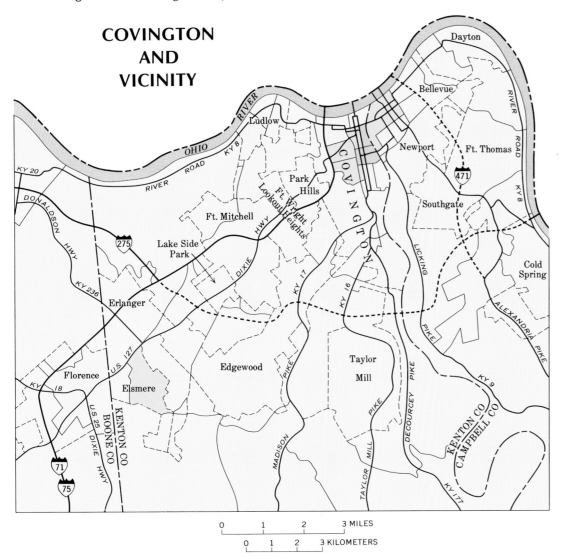

LEXINGTON

IX. MANUFACTURING AND TRADE

MAPS 62-65

MANUFACTURING

In 1974 Kentucky had a labor force of 1,347,000 persons, of whom 292,000 were employed in manufacturing. Employment in Kentucky's manufacturing industries increased rapidly in the 1960s, and from 1970 to 1974 the rate of growth was 10 percent, about four times the national rate.

Employment in manufacturing at Kentucky's principal industrial centers is portrayed on **Map 62**. Metal and machinery industries dominate the manufacturing pattern. Included in this category is a wide variety of products, such as steel, sheet metal, tools, stoves, domestic wares, electrical machinery, transportation equipment and instruments, structural and ornamental steel, metal doors and windows, constructional trim, farm machinery, industrial machinery, and a variety of electrical equipment ranging from vacuum cleaners to television sets. The second most important group of manufacturing industries consists of a variety of chemical and allied industries, including petroleum, coal, rubber, and plastic products. Food processing industries based on farm products such as meat, dairy products, grain products, baked goods, preserved fruits and vegetables, candy, popcorn, flavoring extracts, and syrups, as well as lumber and wood industries, including the manufacture of furniture and fixtures, and stone, clay, and glass products, form a third group of industries located mainly in Louisville, Harrodsburg, and Danville. Manufacturing of tobacco products, the state's fourth major industry, is located in Louisville, Lexington, and Owensboro. The production of distilled liquors, mainly whiskey, is one of Kentucky's larger manufacturing industries, located in the central Bluegrass and in Bardstown and Louisville. In recent

National-Southwire Aluminum Company's plant at Hawesville. Kentucky's aluminum industry, concentrated along the Ohio River at Hawesville, employed 1,356 persons in 1976 and produced such primary products as rolling coils, ingots, wire strand, and rod wire.

MANUFACTURING EMPLOYMENT

MAP 62

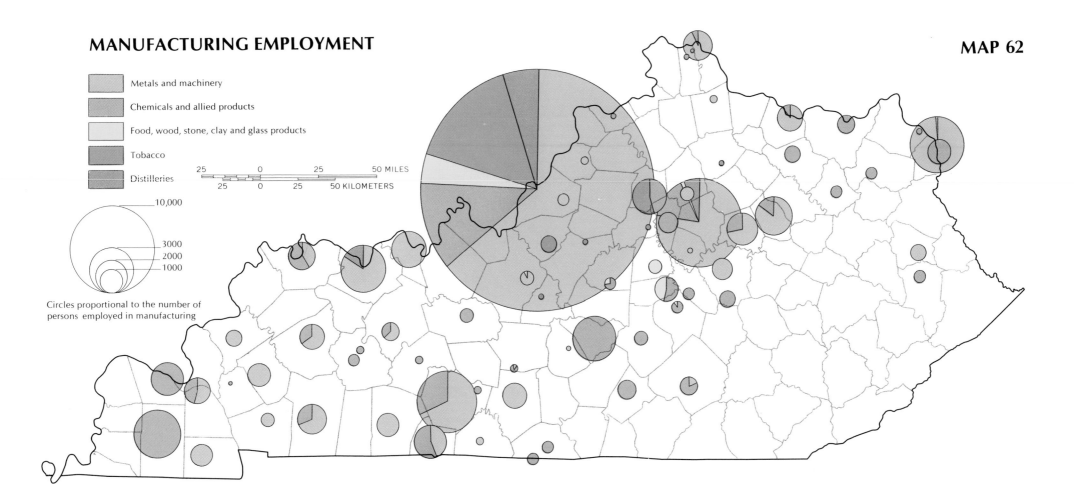

Metals and machinery

Chemicals and allied products

Food, wood, stone, clay and glass products

Tobacco

Distilleries

25   0   25   50 MILES

25   0   25   50 KILOMETERS

10,000

3000
2000
1000

Circles proportional to the number of
persons employed in manufacturing

years employment in all categories of Kentucky industries has increased except for food processing and distilled liquors, where
there has been a decrease.

The largest single employer in 1975 was General Electric, with
approximately 25,000 employees. Forty-eight other manufacturers
in Kentucky employ more than 1,000 persons each and eleven of
them employ more than 2,000. The larger corporations in Kentucky include Ford Motor Company (7,354 employees), International Business Machines (6,493), International Harvester Company
(6,226), Brown and Williamson Tobacco Corporation (5,205),
Northwest Industries (5,050), and Armco Steel Corporation (4,654).
While industries are scattered throughout the state, the principal
areas of growth are the Bluegrass, the Ohio Valley, and western
Kentucky. Southeastern Kentucky has little or no industrial development.

Armco Steel Corporation at Ashland,
employing 4,465 persons in 1976,
produced iron, steel sheets, and coils.
Over 30 percent of the primary metals
employment in Kentucky is in the
Ashland area, dominated by Armco's
large production facility.

MAP 63

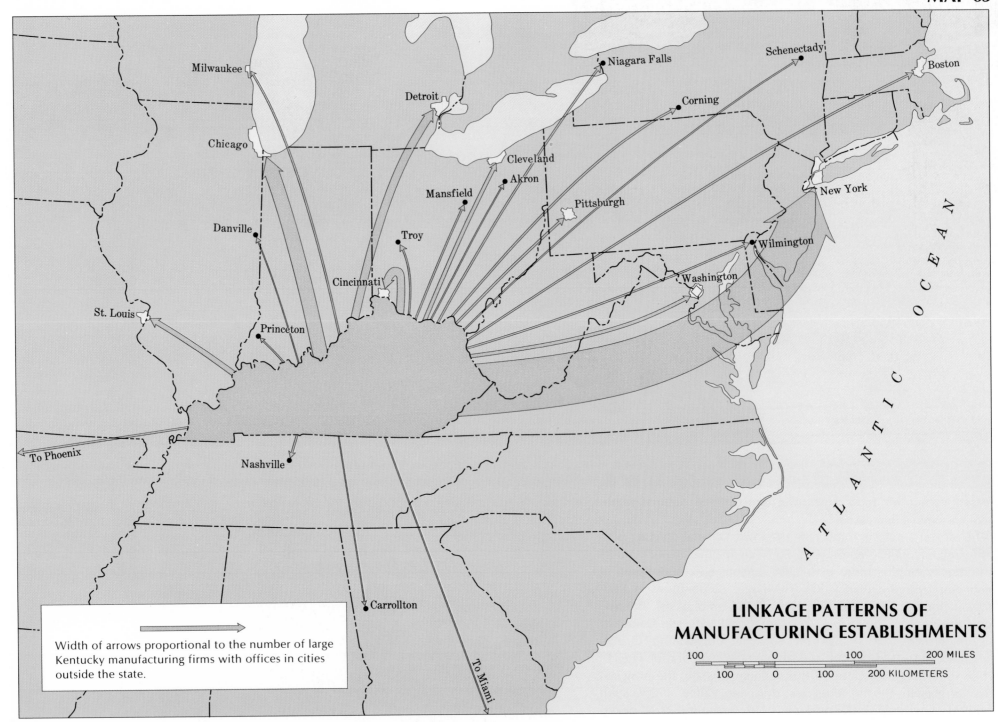

Milwaukee

Detroit

Chicago

Niagara Falls

Schenectady

Boston

Corning

Cleveland

Akron

Mansfield

Pittsburgh

New York

Danville

Troy

Cincinnati

Washington

Wilmington

St. Louis

Princeton

To Phoenix

Nashville

Carrollton

To Miami

ATLANTIC OCEAN

Width of arrows proportional to the number of large Kentucky manufacturing firms with offices in cities outside the state.

LINKAGE PATTERNS OF MANUFACTURING ESTABLISHMENTS

100   0   100   200 MILES

100   0   100   200 KILOMETERS

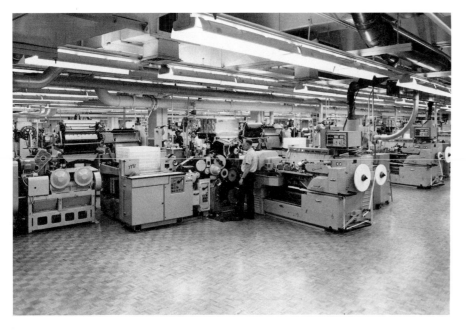

Most of Kentucky's largest industries have linkages outside the state **(Map 63)**. Of the state's sixty-one largest manufacturing industries (those that employ more than 800 persons each), only six have their home offices in Kentucky. Nineteen have home offices in New York, ten in Ohio, and seven in Illinois. Others are in Michigan and Missouri (three each); Connecticut, Indiana, New Jersey, Pennsylvania, and Wisconsin (two each); and Arizona, Delaware, Florida, Georgia, Massachusetts, Tennessee, and Washington, D.C. (one each). New York City dominates all other cities. Kentucky's largest home-based industry is Ashland Oil, Incorporated, located at Ashland.

Several facts stand out in an examination of Kentucky's rapid industrial growth: (1) Metals-related industries provide 47 percent of the state's manufacturing jobs. Kentucky's location at the center of a twenty-eight state area which has 73 percent of the nation's population and 85 percent of the nation's employment in metals promises continued growth in the manufacture of metals-related goods. (2) The state's smaller communities had the sharpest growth in manufacturing in the 1960s, a growth which was facilitated by improvements in highways. (3) Capital investment for new and expanded plants exceeded $3.1 billion during the 1960–1972 period. (4) The average hourly wage for production workers in Kentucky in 1974 was $4.30, or $170.71 per week. (5) Value added was about $6 billion.

Kentucky industry is notable for its variety. *Above left:* Brown and Williamson Tobacco Corporation's plant in Louisville. A British-owned company employing 4,784 persons in 1976, it accounts for 37 percent of Kentucky's tobacco manufacturing employment. One in every five cigarettes produced in the United States is made in Louisville. The Airco Alloy plant at Calvert City (*above*) employed 721 persons in 1976. Calvert City, some twenty miles east of Paducah on the Tennessee River, is the scene of vigorous growth in chemical industries. Much of the supply of raw materials is delivered by barge. *Left:* Firestone Textiles Company at Bowling Green, which had 618 employees in 1976, produces tire fabric. About 7,200 workers are employed by the state's twenty-four textile plants, which also produce knitwear and hosiery.

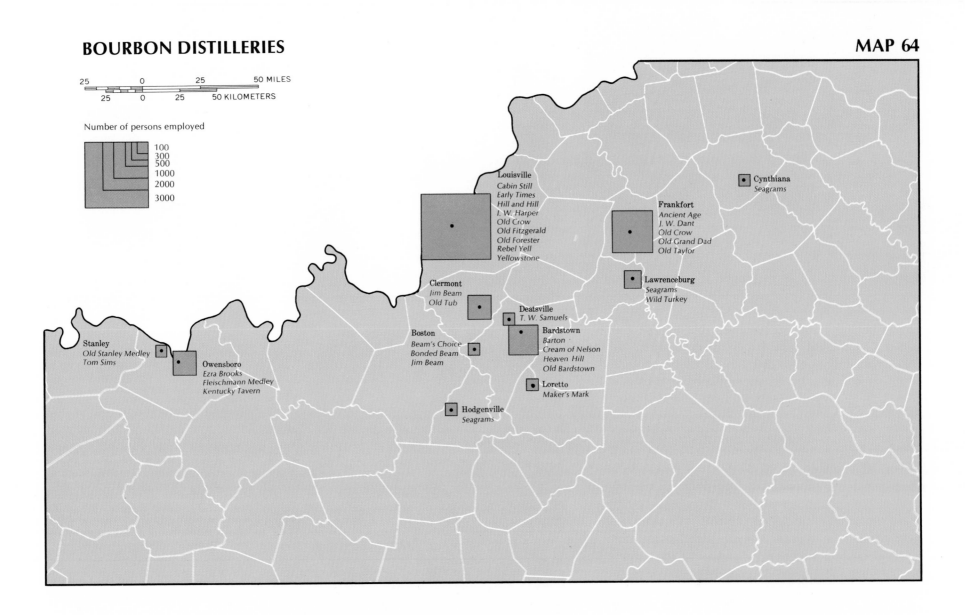

BOURBON DISTILLERIES

MAP 64

25   0   25   50 MILES

25   0   25   50 KILOMETERS

Number of persons employed

100
300
500
1000
2000
3000

Louisville
Cabin Still
Early Times
Hill and Hill
I. W. Harper
Old Crow
Old Fitzgerald
Old Forester
Rebel Yell
Yellowstone

Cynthiana
Seagrams

Frankfort
Ancient Age
J. W. Dant
Old Crow
Old Grand Dad
Old Taylor

Clermont
Jim Beam
Old Tub

Lawrenceburg
Seagrams
Wild Turkey

Deatsville
T. W. Samuels

Boston
Beam's Choice
Bonded Beam
Jim Beam

Bardstown
Barton
Cream of Nelson
Heaven Hill
Old Bardstown

Stanley
Old Stanley Medley
Tom Sims

Owensboro
Ezra Brooks
Fleischmann Medley
Kentucky Tavern

Loretto
Maker's Mark

Hodgenville
Seagrams

The distilled liquors industry has long been an important part of Kentucky's economy. Old Grand-Dad Distillery at Frankfort (*left*), where one of the largest selling brands of bourbon is made, employed 763 persons in 1976. Another 1,500 were employed in the manufacture of white oak barrels (*far left*), mostly in Louisville, Springfield, and Lebanon. Kentucky distilleries now employ about one-third of all the nation's workers in the industry and distill about half of all whiskey produced in the United States.

Kentucky's distilling industry is thought to have been established by German, Irish, and Scottish pioneers from Pennsylvania and other areas who brought with them knowledge of distilling techniques and a palate for the spirituous produce. They found in the Bluegrass everything required for producing high-quality whiskey: nearly mineral-free spring water, ample supplies of high-quality corn and small grains such as rye, and an abundance of white oak for cooperage.

Bourbon whiskey is historically significant in the United States, and especially in Kentucky, because of its long and successful commercial history and its worldwide acknowledgment as the highest form of the American distiller's art. Two factors distinguish bourbon from other distilled spirits. One is that bourbon is distilled from a mash made up primarily of corn, a minimum of 51 percent corn according to federal regulations. The second is that it is aged in white oak barrels that have been charred on the inside and are used only once.

Kentucky's modern bourbon industry is comprised of twenty-seven distilleries, all but four of which are in the Bluegrass (**Map 64**). Each year, Kentucky distilleries purchase over a billion pounds of grain, including 761 million pounds of corn, 161 million pounds of rye, and 122 million pounds of barley malt. Kentucky's present-day production of rye and malting barley is very small and much of the corn crop is fed to livestock. Consequently, the bulk of the distilling grains now comes from the Corn Belt or from Canada.

When the distilling process has been completed the whiskey barrels are stored for four to twelve years in huge warehouses which are closely monitored by the Alcohol, Tobacco, and Firearms Division of the Internal Revenue Service. Whiskey is taxed at the rate of $10.50 per gallon for 100 proof (50 percent alcohol), 90 percent of $10.50 for 90 proof, and so on. Tight security is maintained during the aging and bottling process to assure that all whiskey is accounted for and all taxes are paid. The entire whiskey distilling industry in Kentucky pays approximately $660 million per year in federal excise taxes. Between 75 and 80 percent of the nation's bourbon comes from Kentucky. The industry employs more than 8,000 workers and pays about $38 million a year in state and local taxes.

MAP 65

TOBACCO WAREHOUSES AND MANUFACTURING PLANTS

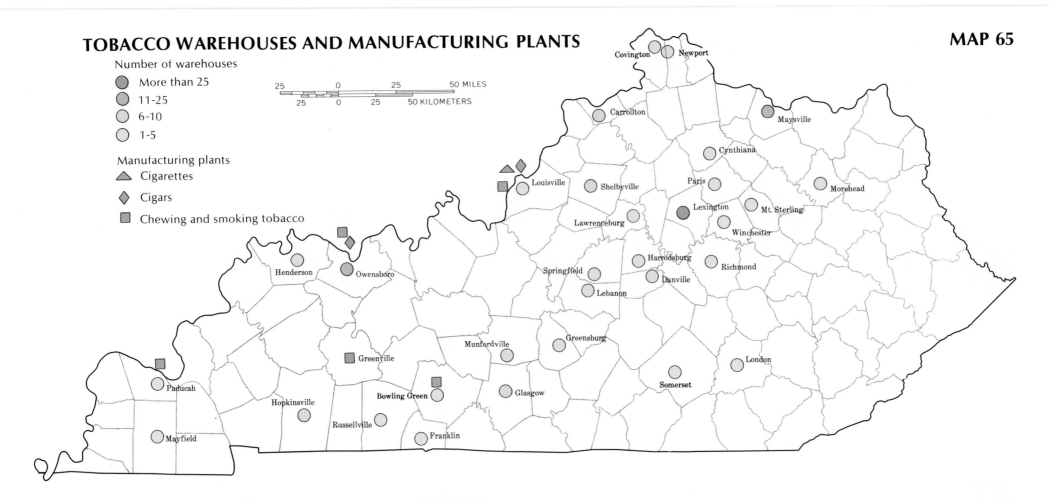

Number of warehouses
- More than 25
- 11-25
- 6-10
- 1-5

Manufacturing plants
- ▲ Cigarettes
- ◆ Cigars
- ■ Chewing and smoking tobacco

25   0   25   50 MILES
25   0   25   50 KILOMETERS

Covington Newport
Carrollton
Maysville
Cynthiana
Louisville Shelbyville Paris Morehead
Lexington Mt. Sterling
Lawrenceburg Winchester
Henderson Owensboro Harrodsburg Richmond
Springfield Danville
Lebanon
Greensburg
Munfordville London
Greenville
Somerset
Paducah Bowling Green Glasgow
Hopkinsville
Russellville
Mayfield Franklin

These farmers are delivering their tobacco to a Lexington tobacco warehouse, where it will be auctioned. The leaves have been stripped and tied around the butts to make small bundles called "hands," which are air cured on the farm before being trucked to the warehouse. In the past few years there have been few changes in the number of tobacco markets, warehouses, and selling periods in Kentucky.

Kentucky has 154 burley tobacco warehouses in 31 cities (**Map 65**). Each year from mid-November to mid-February, burley valued at nearly half a billion dollars is sold at auctions in Kentucky. Tobacco warehouses provide seasonal employment for many people. Lexington, with 26 warehouses, is the world's largest burley market. Kentucky's 1974 burley crop brought a record average of more than a dollar a pound, for a total of $468 million. Dark tobacco is grown in western Kentucky.

Tobacco manufacturing, with principal concentration in Louisville, provides employment for more than 17,000 Kentuckians. Cigarettes are made in Louisville, cigars in Owensboro and Louisville, and chewing and smoking tobacco in Louisville, Owensboro, Paducah, Greenville, and Bowling Green. Twenty-five stemming and redrying plants operate in fourteen Kentucky cities.

WILFORD A. BLADEN

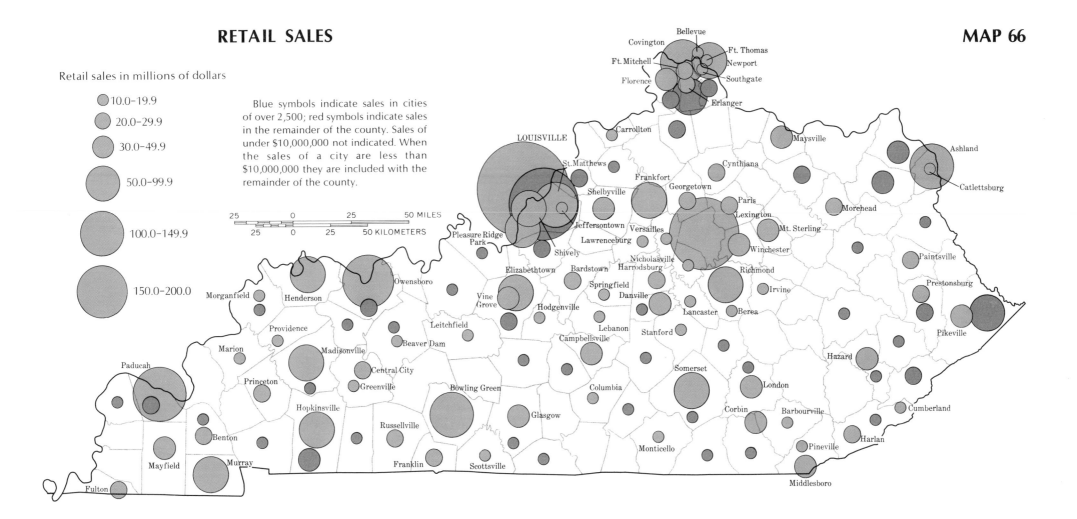

Retail sales in millions of dollars

- 10.0–19.9
- 20.0–29.9
- 30.0–49.9
- 50.0–99.9
- 100.0–149.9
- 150.0–200.0

Blue symbols indicate sales in cities of over 2,500; red symbols indicate sales in the remainder of the county. Sales of under $10,000,000 not indicated. When the sales of a city are less than $10,000,000 they are included with the remainder of the county.

25   0   25   50 MILES
25   0   25   50 KILOMETERS

TRADE

MAPS 66 and 67

Retail trade is a significant function in virtually every city in Kentucky. The cities and towns serve as distributing centers for the areas around them, and the flows of goods and people connect these urban centers to their surrounding hinterlands. Small cities perform only a few functions while larger cities perform many functions. A small town may offer only a general store and a gas station, for example, while a larger town may offer banks and department and specialty stores. The largest cities offer highly specialized goods and services. There is a hierarchical relationship ranging from the smallest town to the largest city. Large cities provide some functions for a wide area, including functions for nearby towns that are lower in the hierarchy.

A good picture of the urban hierarchy of Kentucky is provided by a comparison of total retail sales in various cities **(Map 66)**. Total sales are closely related to the size of population of a city, but significant differences exist between population rank and importance in retail trade **(Table 7)**. The first and second cities of Kentucky, in both population and retail sales, are Louisville and Lexington. Below this level, however, differences in rank are evident. Paducah, which is quite distant from competing major urban centers, has very high retail sales for its population. Saint Matthews, a suburb of Louisville, has very high retail sales for its population because of the presence of several large suburban shopping centers. Many small urban centers seem to be low in retail sales

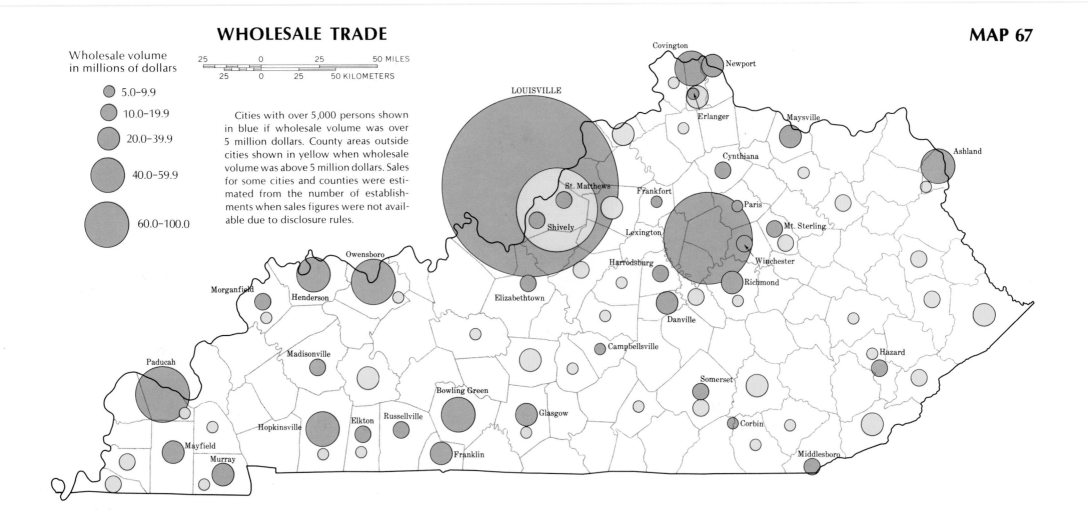

WHOLESALE TRADE

MAP 67

Wholesale volume
in millions of dollars

- 5.0–9.9
- 10.0–19.9
- 20.0–39.9
- 40.0–59.9
- 60.0–100.0

25   0   25   50 MILES
25   0   25   50 KILOMETERS

Cities with over 5,000 persons shown in blue if wholesale volume was over 5 million dollars. County areas outside cities shown in yellow when wholesale volume was above 5 million dollars. Sales for some cities and counties were estimated from the number of establishments when sales figures were not available due to disclosure rules.

Map labels: Covington, Newport, Erlanger, Maysville, LOUISVILLE, Cynthiana, Ashland, St. Matthews, Frankfort, Paris, Shively, Lexington, Mt. Sterling, Owensboro, Harrodsburg, Winchester, Morganfield, Henderson, Elizabethtown, Richmond, Paducah, Madisonville, Danville, Campbellsville, Hazard, Bowling Green, Glasgow, Somerset, Hopkinsville, Elkton, Russellville, Corbin, Mayfield, Franklin, Murray, Middlesboro

because of the proximity of larger centers. For example, cities of the Inner Bluegrass such as Versailles, Georgetown, Nicholasville, and Paris have relatively small amounts of retail sales because of the proximity of Lexington. Covington, Newport, and nearby towns also appear to have retail trade siphoned off by Cincinnati, which is easily accessible across the Ohio River. In the Mountains retail centers are few and small because of sparse population and low incomes.

Wholesale trade **(Map 67)** is much more centralized than retail trade because a few large cities are centers for wide surrounding areas. Louisville is overwhelmingly dominant in wholesale trade, with $1.7 billion of the state's $3.7 billion in wholesale trade in

1967. Jefferson County, including both city and suburban areas, accounted for over $2.0 billion in wholesale trade. Thus while Jefferson County represents only slightly more than one-quarter of the state's retail trade, it represents over half the wholesale trade. Other wholesale centers having $40 million or more in sales in 1967 were Ashland, Bowling Green, Covington, Hopkinsville, Lexington, Owensboro, and Paducah. In 1967 there were 3,715 wholesale establishments in the state, with 903 in Louisville, 229 in Lexington, 126 in Paducah, 92 in Owensboro, and smaller numbers in other cities.

PHILLIP D. PHILLIPS

TABLE 7
Cities with over $40 Million in Retail Sales, 1972

Rank in retail sales	City	1972 sales (in millions of dollars)	1970 population
1	Louisville	891,751	357,911
2	Lexington	504,646	172,456
3	Paducah	152,128	31,627
4	Owensboro	150,122	50,316
5	Covington	144,761	52,482
6	Bowling Green	141,875	36,253
7	St. Matthews	133,705	12,925
8	Ashland	109,347	29,245
9	Newport	86,575	25,974
10	Pleasure Ridge Park	78,385	28,534
11	Hopkinsville	76,261	21,250
12	Frankfort	72,723	21,356
13	Elizabethtown	72,477	11,748
14	Henderson	67,234	22,939
15	Somerset	66,795	10,436
16	Madisonville	66,596	15,332
17	Shively	66,210	19,304
18	Murray	58,638	13,537
19	Richmond	56,201	16,861
20	Glasgow	48,654	11,301
21	Florence	47,452	11,457
22	Mayfield	46,631	10,724
23	Winchester	46,000[a]	13,402
24	Middlesboro	43,692	12,124

[a]Sales estimates are based on the number of establishments in the city and county; actual sales have not been released because of disclosure rules.

The single service country store selling gasoline and groceries, such as the Fayette County general store illustrated (*top right*), serves local residents in its rural area. It represents the lowest tier in the commercial hierarchy. Hyden (*above*), seat of Leslie County, is an example of an intermediate retail service center with stores and other firms serving county residents. Large cities stand at the top of the commercial hierarchy, offering a wide range of goods and services and serving as regional retail sales centers. *Above left:* Shopping in a Lexington department store. Lexington's retail sales area covers most of central Kentucky.

Some of Kentucky's more interesting geologic features. *Above left:* One entrance to Mammoth Cave, located in the Mississippian limestone area near Cave City. One of the world's great natural caverns, with over 180 miles of passageways, it is the product of geologic action and erosion over many millions of years. Cumberland Falls (*left*), on the Cumberland River near Corbin, is the largest waterfall east of the Rockies and south of Niagara. It results from interruption of valley deepening by the outcropping Rockcastle conglomerate. By night at full moon the falls show a mysterious moonbow, one of two such phenomena on earth. Natural Bridge (*above*), near Slade in Powell County, was formed in the Rockcastle sandstone and conglomerate belt near the western margin of the Cumberland Plateau. It is the result of disintegrating action by wind, rain, and frost on softer conglomerate limestone under the hard capstone.

X. GEOLOGY, MINERALS, AND ENERGY

GEOLOGY AND TECTONICS

Kentucky's natural regions are directly related to the underlying rock strata **(Map 68)**. Most of the formations are sedimentary, formed near the end of the Paleozoic Era some 200 million years ago. The oldest exposed rocks are limestone of Ordovician age. They contain a few layers of shale and siltstone and form the surface of the Bluegrass Region in central Kentucky. The youngest Paleozoic strata exposed in the state are Pennsylvanian rocks, which consist mainly of sandstones, conglomerates, shale, and coal. Pennsylvanian rocks are found at the surface in both the Western Coal Field and the eastern Mountains and coal field. Surface rocks in the Pennyroyal are of Mississippian age, mainly limestone but with some shales, siltstone, and sandstones. The Mississippian rocks are older than those of Pennsylvanian age but younger than the limestones, shales, and sandstones of the Devonian and Silurian ages which are layered over the Ordovician rocks. The Devonian and Silurian rocks are exposed in the Knobs surrounding the Bluegrass. The Jackson Purchase is underlain by much younger rocks formed when the area was covered by the Gulf of Mexico.

Central Kentucky has been pushed up by geologic forces; the main axis of uplift, the Cincinnati Arch, crosses the state in a northeast–southwest direction from Lexington toward Nashville, Tennessee **(Map 69)**. More recent rocks have been eroded from the crest of the arch, leaving older Ordovician rocks exposed at the surface in the Bluegrass. Around the Bluegrass the rocks are pro- gressively younger, so that the geologic map resembles an onion with one side sliced off, exposing the layers.

In easternmost Kentucky the land has been subjected to folding and faulting, resulting in the uplift of the Pine and Cumberland mountain ranges. The highest elevations in the state are here, and most of the state's drainage is consequently from the southeast toward the northwest. The eastern Kentucky Mountains include the Cumberland Plateau—Pine and Cumberland mountains and the Middlesboro syncline (see geological cross section CC'). It is a maturely dissected area where local variation in altitude and relief is the expression of variation in rock outcrop and character.

The land surface features of Kentucky are largely a result of the kinds of rocks beneath the surface. Most of the area underlain by sandstones is either hilly or mountainous because sandstones tend to resist weathering and erosion. The plains in Kentucky are generally associated with limestone, which erodes more easily, permitting underground drainage, sinks, and caves. High cliffs and knobs form the boundaries of the various regions where different rock strata are juxtaposed. Many natural bridges, waterfalls, and other spectacular features are found near the western edge of the Cumberland Plateau, while the river bottomlands in the extreme west are wide, flat, and swampy.

WILFORD A. BLADEN

GENERALIZED GEOLOGY

MAP 68

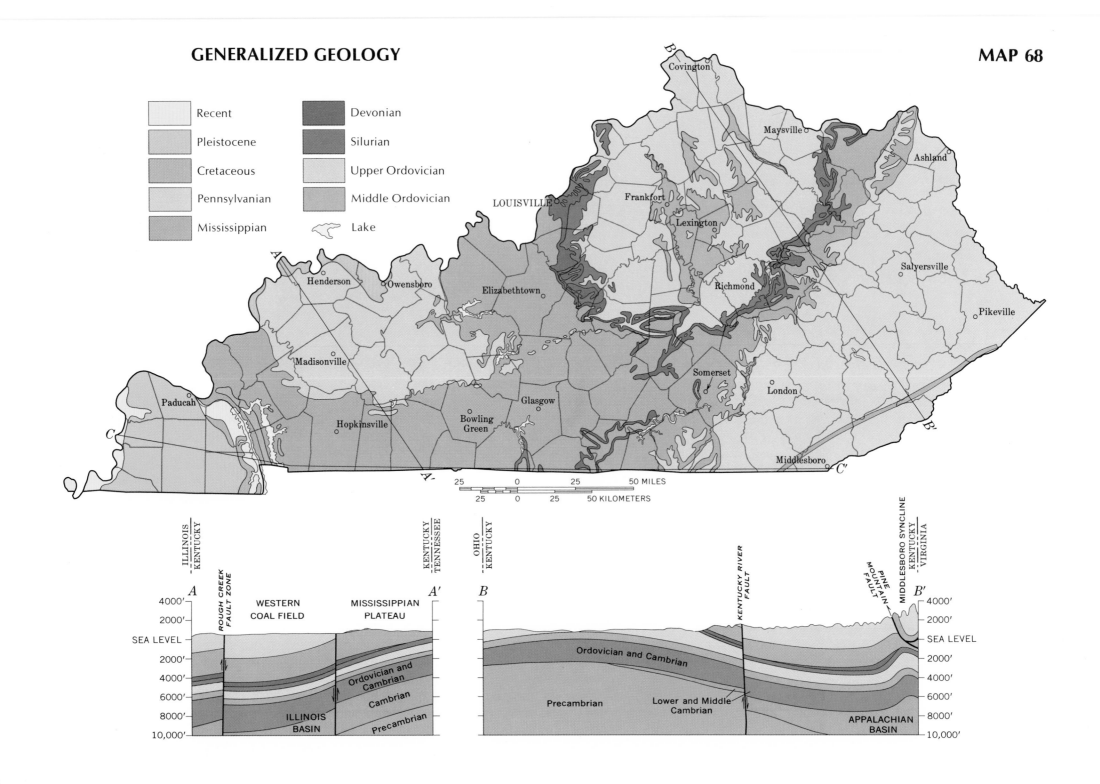

Recent
Pleistocene
Cretaceous
Pennsylvanian
Mississippian

Devonian
Silurian
Upper Ordovician
Middle Ordovician
Lake

Covington
Maysville
Ashland
LOUISVILLE
Frankfort
Lexington
Salyersville
Henderson
Owensboro
Elizabethtown
Richmond
Pikeville
Madisonville
Somerset
London
Paducah
Glasgow
Hopkinsville
Bowling
Green
Middlesboro

25   0   25   50 MILES
25   0   25   50 KILOMETERS

ILLINOIS
KENTUCKY
KENTUCKY
TENNESSEE
OHIO
KENTUCKY
KENTUCKY
VIRGINIA

A
ROUGH CREEK FAULT ZONE
WESTERN COAL FIELD
MISSISSIPPIAN PLATEAU
A'
B
KENTUCKY RIVER FAULT
PINE MOUNTAIN FAULT
MIDDLESBORO SYNCLINE
B'

4000'
2000'
SEA LEVEL
2000'
4000'
6000'
8000'
10,000'

Ordovician and Cambrian
Cambrian
ILLINOIS BASIN
Precambrian

Ordovician and Cambrian
Precambrian
Lower and Middle Cambrian
APPALACHIAN BASIN

4000'
2000'
SEA LEVEL
2000'
4000'
6000'
8000'
10,000'

98

TECTONICS

MAP 69

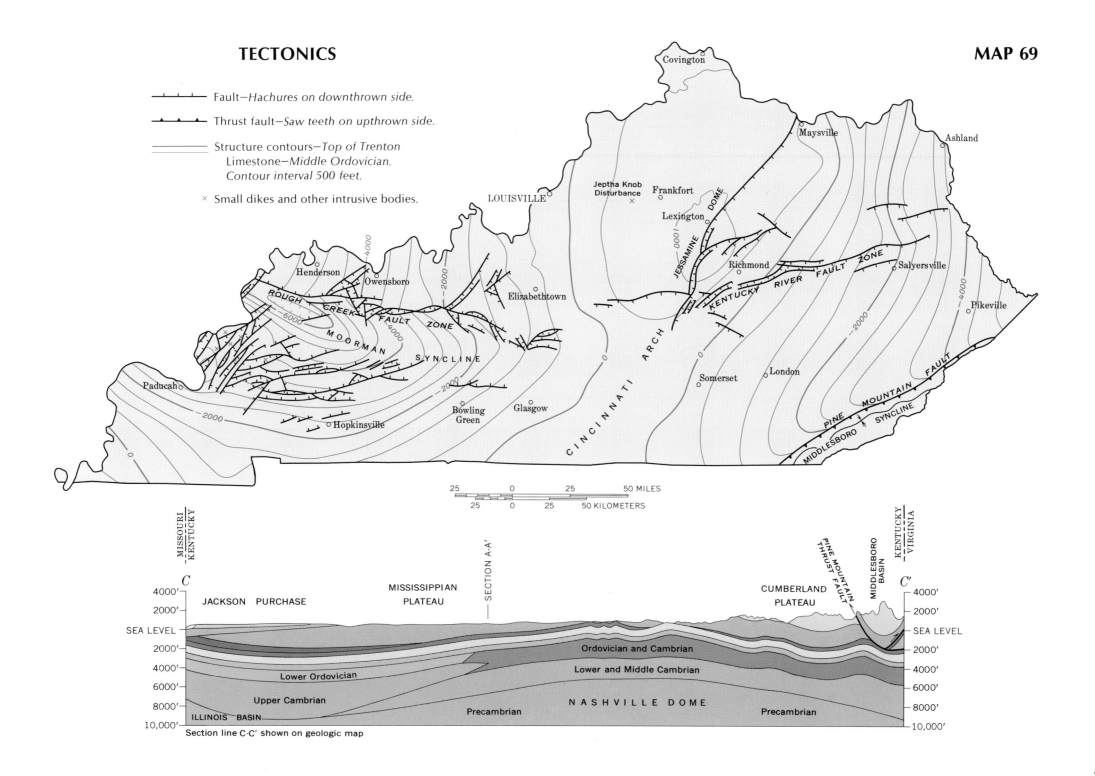

Fault—*Hachures on downthrown side.*

Thrust fault—*Saw teeth on upthrown side.*

Structure contours—*Top of Trenton Limestone—Middle Ordovician. Contour interval 500 feet.*

× Small dikes and other intrusive bodies.

Covington

Maysville

Ashland

Jeptha Knob Disturbance

Frankfort

LOUISVILLE

Lexington

JESSAMINE DOME

Richmond

KENTUCKY RIVER FAULT ZONE

Salyersville

Pikeville

Henderson

Owensboro

ROUGH CREEK FAULT ZONE

Elizabethtown

MOORMAN SYNCLINE

CINCINNATI ARCH

Somerset

London

Paducah

Hopkinsville

Bowling Green

Glasgow

PINE MOUNTAIN FAULT

MIDDLESBORO SYNCLINE

25 0 25 50 MILES

25 0 25 50 KILOMETERS

MISSOURI / KENTUCKY

C

4000'

2000'

SEA LEVEL

2000'

4000'

6000'

8000'

10,000'

JACKSON PURCHASE

MISSISSIPPIAN PLATEAU

SECTION A-A'

Lower Ordovician

Upper Cambrian

ILLINOIS BASIN

Ordovician and Cambrian

Lower and Middle Cambrian

Precambrian

NASHVILLE DOME

CUMBERLAND PLATEAU

PINE MOUNTAIN THRUST FAULT

MIDDLESBORO BASIN

KENTUCKY / VIRGINIA

C'

4000'

2000'

SEA LEVEL

2000'

4000'

6000'

8000'

Precambrian

10,000'

Section line C-C' shown on geologic map

MINERAL RESOURCES AND INDUSTRIES

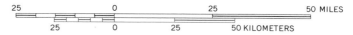

25   0   25   50 MILES

25   0   25   50 KILOMETERS

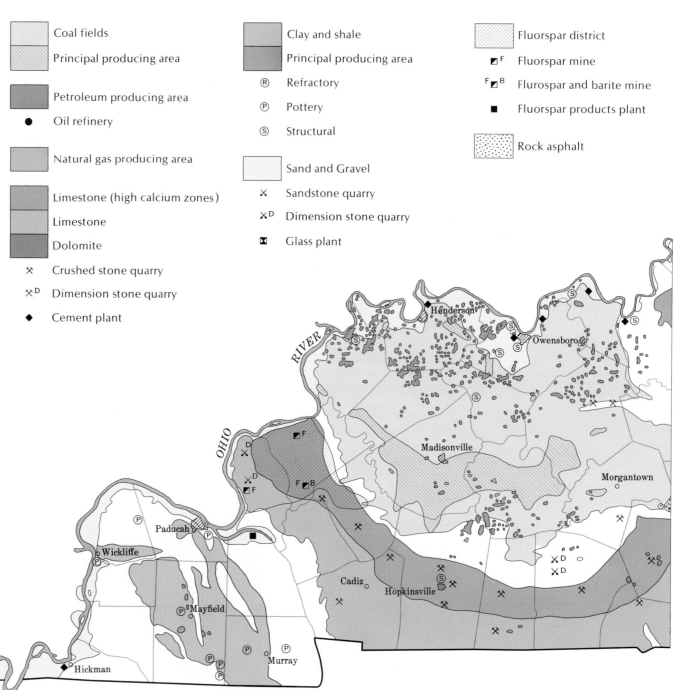

Coal fields

Principal producing area

Petroleum producing area

● Oil refinery

Natural gas producing area

Limestone (high calcium zones)

Limestone

Dolomite

✕ Crushed stone quarry

✕D Dimension stone quarry

◆ Cement plant

Clay and shale

Principal producing area

Ⓡ Refractory

Ⓟ Pottery

Ⓢ Structural

Sand and Gravel

✕ Sandstone quarry

✕D Dimension stone quarry

▣ Glass plant

Fluorspar district

◪F Fluorspar mine

F◪B Flurospar and barite mine

■ Fluorspar products plant

Rock asphalt

MAP 70

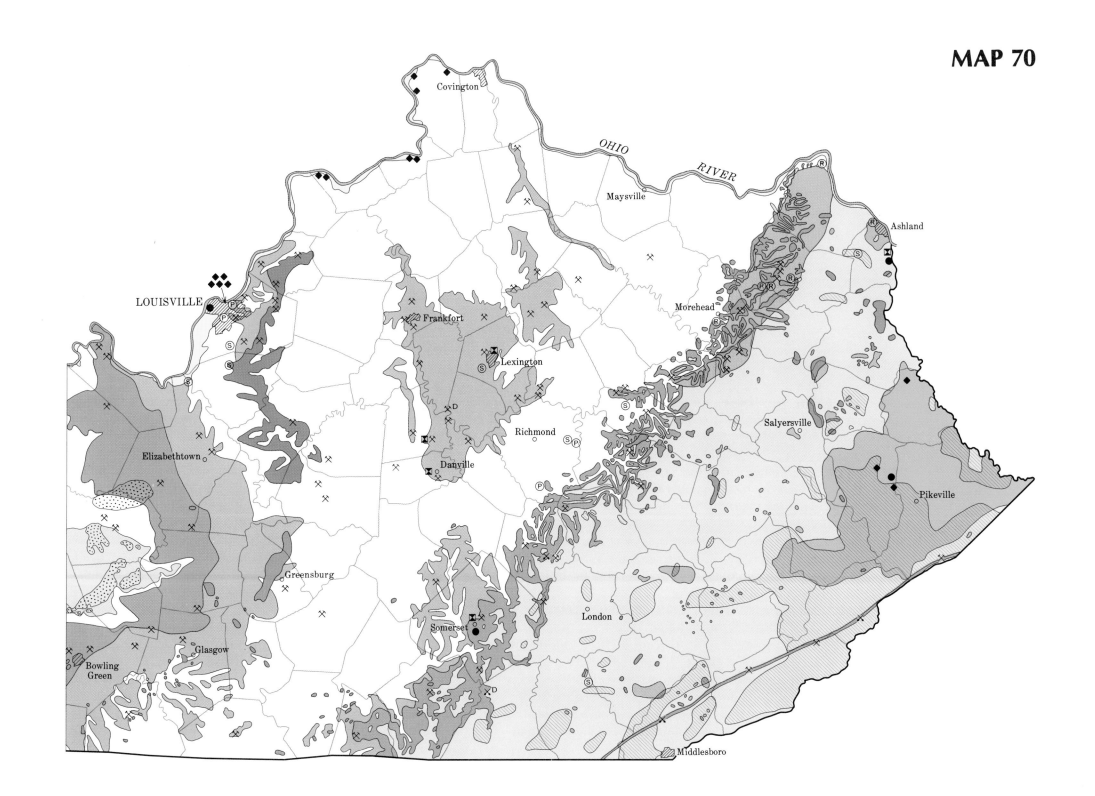

Covington

OHIO

RIVER

Maysville

Ashland

LOUISVILLE

Morehead

Frankfort

Lexington

Salyersville

Richmond

Elizabethtown

Danville

Pikeville

Greensburg

London

Somerset

Glasgow

Bowling
Green

Middlesboro

MINERAL RESOURCES

Kentucky's mineral resources (Map 70) are a vital element in the economy of the state. Coal mining alone provided jobs for more than 38,000 persons and incomes of more than one billion dollars annually in 1975. Kentucky ranks first in the nation in annual bituminous coal production, with more than 140 million tons produced in 1975. Reserves are estimated at 66 billion tons. Both underground and surface mines are operated in the eastern and western coal fields, although underground mines produce more coal in the eastern field, while strip mining predominates in the western. Slightly more than 60 percent of the state's annual coal production comes from the eastern field. Kentucky had more than 1,498 coal mines in 1973, distributed in forty counties.

Kentucky's production of oil and gas is not very large and does not employ as many people as the coal industry. The largest producing oil field is the Big Sinking Creek Field near Beattyville in Lee County, while the largest gas field is in the northeast, near Ashland. More than half of Kentucky's counties produce some oil and/or gas. Although oil wells are widely distributed in eastern and western Kentucky, no oil has been discovered in the Bluegrass. Kentucky oil wells produced less than nine million barrels in 1973 and production is declining. Kentucky natural gas wells in 1974 produced 58.5 billion cubic feet and supplied about 30 percent of the state's needs; the remainder comes mostly from fields in Texas and Louisiana. Production in Kentucky is declining, however.

Other minerals contribute significantly to the state's economy, including limestone, sand, gravel, fluorspar, and a variety of clays. Zinc, lead, and asphalt are obtained in small quantities, usually as by-products of other mining operations.

Of all the minerals, coal is the state's most important natural resource. Kentucky's share of total United States coal production rose from 16 percent in 1960 to 23 percent in 1974. During the 1950s oil and natural gas were often substituted for coal. Today's energy shortage, however, may return coal, the fuel with the largest domestic reserves, to its former position of dominance among all fuels. Recently the demand and price for coal have soared. The result has been an increase in coal mining, with many coal counties gaining in employment and wages from the industry.

Kentucky is the only coal-producing state with two major coal fields. The eastern field is part of the Appalachian Coal Region.

The Kentucky portion of this region includes all or parts of thirty-one counties. The Western Coal Field includes eighteen Kentucky counties. They form the southeastern end of the Eastern Interior Coal Region, which lies mainly in Kentucky, Indiana, and Illinois, with small parts in Iowa and Missouri.

The coal fields in the eastern and western parts of the state lie in two distinct geographical areas with differences in both the nature of coal deposits and mining practices. In the eastern coal field the seams are relatively thin and the topography is rugged. Many small or medium-size truck mines (surface, underground, or auger) and a relatively few major underground and surface mines characterize the region. The coal is obtained by removing the overburden and scooping up the coal, or by augering or tunnelling into the seam from the outcropping. The western field, on the other hand, is characterized by thick seams and gently rolling topography. The coal lies below the surface in horizontal seams and is obtained by stripping or by sinking a shaft to the seam level. Mining is mainly by surface methods and mines are generally very large compared to those in eastern Kentucky. In both eastern and western Kentucky over 50 percent of the 1974 production, approximately 85 and 52 million tons respectively, was from surface mines.

In recent years Kentucky has had a dramatic increase in the use of surface mining techniques. Coal mined in 1960 by surface techniques represented about one-third of all coal mined in the state; by 1974 over half the state's coal was surface mined. Improved technology, lower labor costs, lower mining costs, and less danger of accidents to miners have made surface mining very attractive to the industry. Surface mining has had an environmental impact where poor mining and reclamation techniques have been used. The environmental dangers are most severe on steeply sloped lands in the eastern coal field. Contemplated federal legislation imposing slope restrictions for surface mining could have a considerable effect in eastern Kentucky.

The different characteristics of coal deposits in the two fields also result in different patterns of marketing and usage. Over 91 percent of western Kentucky coal is used by electric utilities because of its higher sulfur content. Easier access to low-cost water transportation makes it possible to ship western Kentucky coal to markets in Florida, Minnesota, and Pittsburgh, and to export markets. Also, the larger size of the western mines facilitates market-

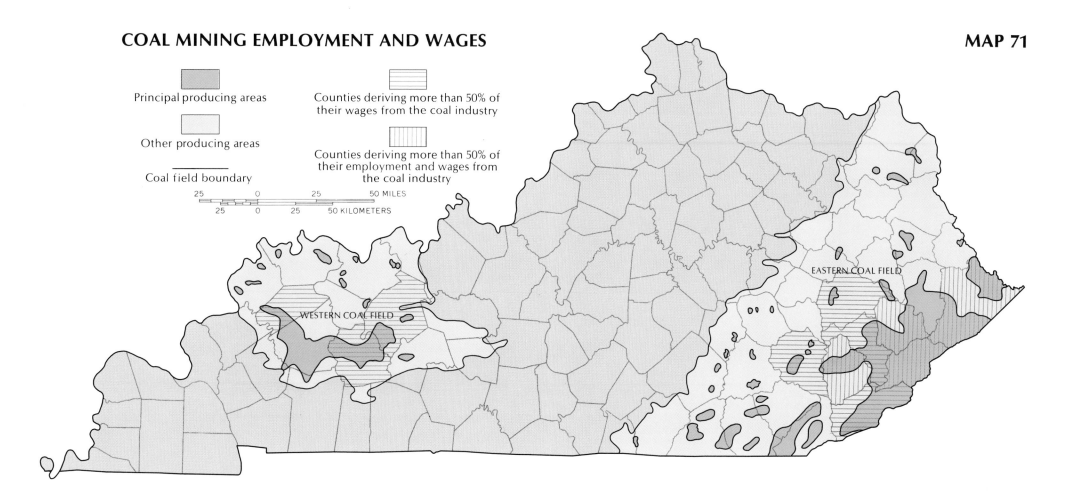

Principal producing areas

Other producing areas

Coal field boundary

Counties deriving more than 50% of
their wages from the coal industry

Counties deriving more than 50% of
their employment and wages from
the coal industry

25   0   25   50 MILES

25   0   25   50 KILOMETERS

WESTERN COAL FIELD

EASTERN COAL FIELD

ing to the high-volume consumers, such as the TVA and Kentucky Utilities Company power plants. The rest goes by rail north to the port of Chicago for transit along the Great Lakes waterway to Ohio, or south by way of the Green, Ohio, and Tennessee rivers to Florence, Alabama, to be loaded on barges for points farther south. Western Kentucky mines have direct mine-to-barge or mine-to-rail facilities, and truck hauling of coal is less important than in the eastern coal field.

Eastern Kentucky coal is generally lower in sulfur content and consequently has more varied uses and a wider distribution pattern. In recent years the low sulfur content of the eastern coal has also increased its demand by electric utilities companies which are trying to meet ambient air standards. Eastern coal is hauled by trucks onto the rail lines at numerous loading points or tipples. A number of larger mines are directly served by a rail spur, but the

coal from the smaller mines is trucked to a nearby tipple, sometimes as much as twenty-five miles away. The large number of small to medium-size mines results in a maze of rail lines and spurs dotted with loading points. Eastern Kentucky coal moves by rail to the Tidewater area for export and to power plants in Virginia, North Carolina, Georgia, and Tennessee.

The influence of coal is very strong in many Kentucky counties **(Map 71)**. For all Kentucky coal counties with 1970 data available, coal mining provided 17 percent of total employment and 23 percent of all wages. In four counties—Pike, Knott, Letcher, and Leslie —coal mining represented more than half of both employment and wages. In Breathitt, Perry, Clay, Harlan, Webster, Muhlenberg, and Ohio counties more than half of all wages were from the coal industry. In 1975, the coal mining industry employed 38,000 people in Kentucky.

SURFACE MINES IN EASTERN KENTUCKY

MAP 72

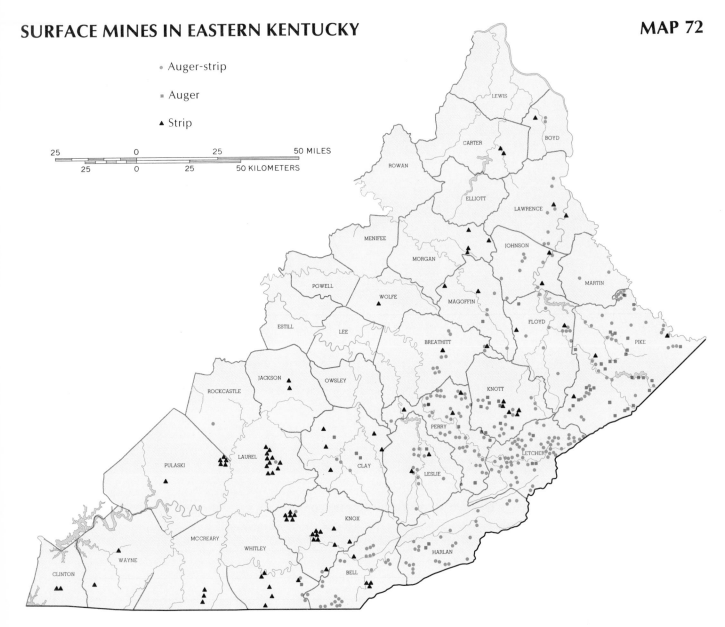

- Auger-strip
- Auger
- ▲ Strip

Surface coal mining in Perry County. This huge power shovel removes soil and rock, then scoops up many tons of coal at a time. Surface mining has increased substantially in the last few years.

The location of surface mines and especially concentrations of surface mines in an area can indicate possible dangers from water pollution. The greatest concentration of surface mining activity is in the Kentucky River Basin, and there is also considerable surface mining in the Big Sandy and Cumberland river basins (**Map 72**). Unless care is taken in the mining process, acid mine drainage can pollute streams in the vicinity of, and in some cases well downstream from, the polluting source. On a county-by-county basis, Letcher, Perry, and Knott counties have the largest number of surface mines in the eastern coal field. These three counties are all within the Kentucky River Basin. (For more detail on pollution from mining see **Map 123**.)

Most of the surface mines in western Kentucky are concentrated in the crescent-shaped principal producing area outlined on **Map 71**. Within this area the surface mining activity is most intense in Hopkins and Muhlenberg counties, where large interhighway areas are being strip-mined.

WILFORD A. BLADEN and RICHARD I. TOWBER

Kentucky's electric generating industry, based largely on coal, is growing at a rate of 9–10 percent annually. The steam generating plant at Paradise in the Western Coal Field (*far left*) is one of the state's largest, with a capacity of 2,000 megawatts. The hydroelectric and steam plant at Dix Dam (*left*), with a capacity of 24 megawatts, is one of many small plants in the state.

ENERGY NETWORKS

MAPS 73 and 74

The electric power used in Kentucky in 1975 was produced from fifty-one generating plants located within the state and from some plants located in adjoining states. More than 90 percent of the electric power generated within Kentucky is from coal-fueled plants, and this proportion is expected to rise even higher in the future.

Of the fifty-one generating plants, thirty-three are steam generator plants, nine are hydroelectric plants, and nine are internal combustion plants. A number of smaller emergency or standby plants which burn oil or natural gas are not included in this count.

In spatial terms, more than half of the electric power plants occupy river-oriented sites in the northern part of the state **(Map 73)**. This northern river zone has thirty-three generating plants, only two of which are hydroelectric. The second largest cluster is the group of three steam, two hydroelectric, and two internal combustion plants in the central Bluegrass Region. A third group of five generating plants, almost evenly spaced, serves south-central and southeastern Kentucky. Six additional plants are concentrated in west-central and western Kentucky.

Currently production of electric energy is carried on within Kentucky by four major (more than 1,000-megawatt capacity) and six smaller electric systems. These ten companies, which both produce and distribute electric power, are: Louisville Gas and Electric Company (LG & E), Kentucky Power Company (KPC), Tennessee Valley Authority (TVA), Kentucky Utilities (KU), Eastern Kentucky Power Cooperative (EKPC), Big Rivers Cooperative (BR), and Henderson Municipal, Owensboro Municipal, Paris Power, and Union Light, Heat and Power companies. In addition, there are fifty-five utility companies engaged solely in the distribution of power, including twenty-seven rural electric cooperatives and twenty-eight municipal distribution systems. Some of the larger electric distributors are: Electric Energy, Inc. (EEI), Public Service Indiana (PSI), Ohio Valley Electric Company (OVEC), Cincinnati Gas and Electric Company (CG & E), Indiana and Michigan Electric Company (I & MEC), and Dayton Power and Light Company (DP & LC). The distribution companies purchase power from the generating companies, which in turn buy and sell power among themselves. A con-

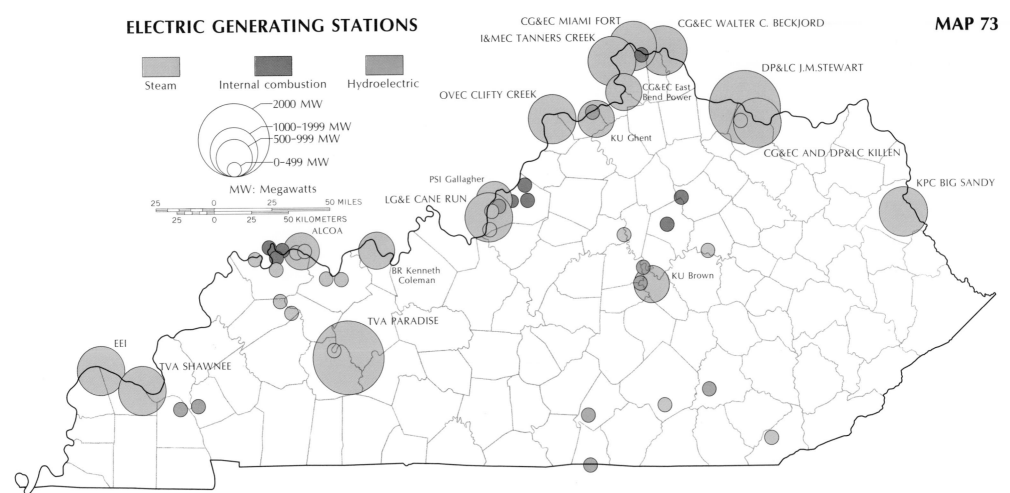

siderable amount of electric power is networked or transferred among major utility companies.

Kentucky's pattern of electrical energy transmission is comprised of a series of nodes, located at generating plants, linked by long-distance transmission lines. The transmission lines linking the generating sites with the consuming areas carry widely varying high voltages, ranging from 34,500 volts to as much as 765,000 volts. Most of the line mileage, however, carries between 69,000 and 161,000 volts. The higher voltage lines are mainly in northern Kentucky. One transmission line of 765,000 volts provides service in part of Kentucky's eastern coal field; one line of 500,000 volts is a TVA line from Tennessee to the Muhlenberg generation station southeast of Central City in Kentucky's Western Coal Field.

The evolution toward larger and larger electric generating plants located farther and farther from power markets has forced the electric utility industry to move steadily toward longer-distance transmission. About 99 percent of all transmission lines are overhead, carried on towers of varying height and requiring substantial right-of-way, on the order of twelve acres per mile. The land cost for transmission lines which cross densely populated areas can be prohibitive. As the need for electric power increases, so will the need for transmission lines and hence more and more land. Underground installation of transmission lines, six to twenty times more expensive than that of overhead lines, is an attractive prospect in terms of environmental aesthetics and land acquisition costs.

Petroleum products account for about 31 percent of Kentucky's total energy consumption. Of the total petroleum consumed in Kentucky in 1974, about 13 percent originated from Kentucky oil fields and the remainder was pumped from wells in other states. A small quantity—less than 1 percent of the total consumed—orig-

MAP 74

OIL AND GAS

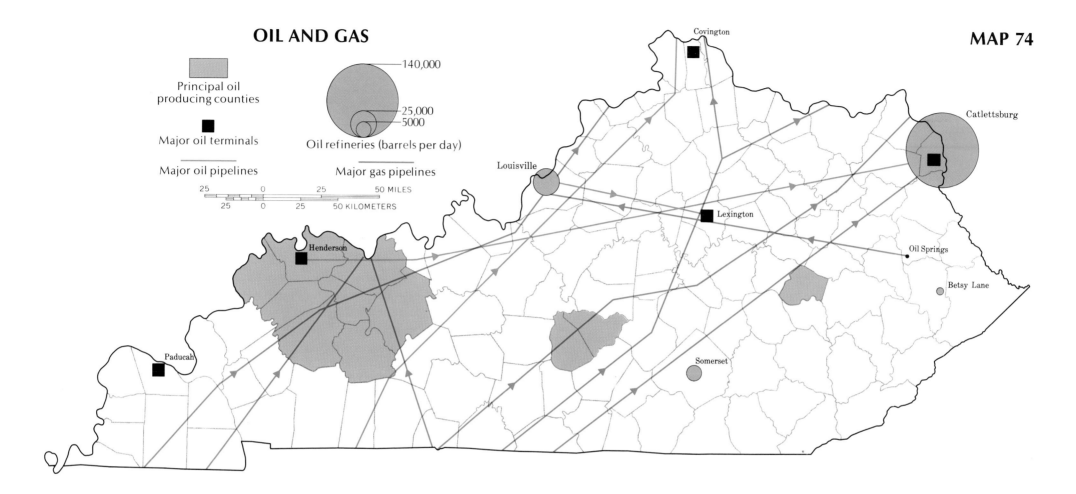

Principal oil producing counties

Major oil terminals

Major oil pipelines

Oil refineries (barrels per day)

140,000

25,000
5000

Major gas pipelines

25   0   25   50 MILES

25   0   25   50 KILOMETERS

inated overseas. **Map 74** shows the locations of oil refineries, major oil pipelines, and major oil terminals in Kentucky, as well as the principal oil-producing counties.

The location of refineries is very sensitive to the location of crude oil sources on the one hand and to markets for finished products on the other. The refinery location must minimize the transportation costs of both input and output. Another locational factor is the prevailing direction of network flow. For the region of which Kentucky is a part, network flow is generally northeast and north from the large crude-producing areas in the southwestern states. For competitive economic reasons, a refinery can afford to ship products farther north and east in the direction of flow; it may not be able to backhaul products very far against the flow. The Ashland Oil Refinery at Catlettsburg ships primarily to the north and east. The three smaller refineries at Louisville, Betsy Layne, and

Somerset ship most of their output to local markets. The combined daily output of these Kentucky refineries (172,000 barrels) is almost identical with the daily consumption of petroleum production in the state, but a substantial portion of this output is not consumed in Kentucky. The state's petroleum product needs are instead largely met by the output of refineries to the south and west.

Kentucky refineries obtain out-of-state crude supplies by barge and in-state crude supplies by local and regional pipelines and some barging. In general, petroleum products distributed in Kentucky are moved by truck to user, distributor, or retailer from bulk terminal points located in and near the state. These terminals are supplied with bulk products by barge and pipeline from various refineries. As with most other energy networks, the movement of petroleum products from source to user is very complex, involving large numbers of modal changes and storage points. Network flow

Ashland Oil Company's refinery at Catlettsburg, where the Big Sandy joins the Ohio River. Petroleum refining is a relatively small industry in Kentucky, providing employment for 1,098 persons in 1976. Petroleum products, used mostly in the transportation sector, account for about 31 percent of the energy consumed in the state. Kentucky industry operates mainly on coal.

rates are affected by several variables, such as seasonal changes of climate and daylight hours, production changes, user demand, technology of end-point converters, fluctuations in other energy networks, price and cost, and industry-government decisions. Kentucky's proximity to major southwest-northeast flow channels for crude oil and petroleum products (the Mississippi and Ohio rivers and pipelines in Indiana and Ohio) makes it possible for the petroleum refinery industry in Kentucky to expand.

Natural gas accounts for about 24 percent of Kentucky's total energy consumption. Of the total natural gas consumed in Kentucky in 1974, the equivalent of about 30 percent originated from Kentucky gas and oil fields and the remainder from elsewhere in the nation, primarily the Gulf Coast states. Most of the natural gas wells are located in oil-producing counties in eastern Kentucky and in the central portion of western Kentucky. Very little Kentucky-

produced gas was consumed in Kentucky. Since the prevailing flow of natural gas in Kentucky is southwest to northeast, and since the bulk of Kentucky's gas production occurs in the easternmost portion of the state, most of Kentucky's natural gas flows northeast into other states.

Kentucky has a major role in natural gas transmission. Eight companies operate interstate transmission lines traversing Kentucky. Of these companies, five (Texas Gas, Tennessee Gas, Columbia Gulf, Texas Eastern, and Columbia Gas) furnish Kentucky distributors. The other three (Trunkline Gas, Midwestern Gas, and Michigan Wisconsin Pipeline) do not serve state distributors. These transmission lines connect major gas-producing areas in the Gulf Coast region with major gas consumption centers in the North and Northeast. **Map 74** shows the natural gas pipeline flow. Kentucky's net consumption of gas amounts to less than 5 percent of the total quantity of gas which flows through the Commonwealth.

The geography of major pipeline transmission has been a boon to Kentucky, permitting the establishment of natural gas distributing systems in a large number of counties located on or near these pipelines. Under conditions of gas scarcity, however, the benefits of proximity to pipelines quickly fade because priority must be given to major customers downstream. Gas, the cleanest and most efficient fuel for many purposes, is not available in sufficient quantity to meet Kentucky's growing demands, despite the fact that huge volumes flow annually through the state.

In addition to the major transmission companies Kentucky is served by four major and forty-two smaller gas distribution companies. The major distributors are: Columbia Gas of Kentucky, serving thirty-seven communities in central and eastern Kentucky; Louisville Gas and Electric, serving Jefferson and seven surrounding counties; Union Light, Heat and Power, serving Campbell, Kenton, Boone, Grant, and Owen counties; and Western Kentucky Gas Company, serving 105 communities in thirty-six central and western counties. The major distribution companies and most of the smaller utility companies purchase gas on long-term contract from the gas transmission companies. Supplies for new gas users in Kentucky cannot be provided by the major transmission companies. Coal gasification, however, has recently come as a potential replacement source for dwindling supplies of natural gas. In Kentucky, private interests and state and federal governments are joining forces to develop, test, and commercially install gasification facilities.

WILFORD A. BLADEN and WILLIAM A. WITHINGTON

XI. LAND USE AND PHYSICAL CHARACTERISTICS

LAND USE

MAPS 75-78

The present land use characteristics of the 25,510,881 acres comprising the state of Kentucky were inventoried by the state's Soil and Water Conservation Needs Committee in 1970. The number of acres in cropland, pasture, forest, and other similar uses totals 23,507,491, or 92.2 percent of the state's land area. The remaining 2,003,430 acres, or 7.8 percent, are classified as noninventory acreage. Lands falling into the latter classification include federally owned noncropland (1,047,416 acres), urban built-up areas (834,858 acres), and small water areas (121,156 acres).* **Map 75** shows land use within the five major physiographic regions of the state, which are delineated on **Map 76**.

Forests and woodlands cover nearly 11 million acres, or 43 percent of Kentucky's total acreage. Over 90 percent of the forested acres are privately owned, divided equally between commercial interests and farmers. Federal and state ownership accounts for the remainder. Although forests are most prevalent in the eastern and south-central portions of the state, all 120 counties in the state have woodlands. Ninety counties are important from a forestry standpoint but the largest percentage of woodlands is in the Mountains. The Bluegrass has the highest proportion of pastureland. Cropland occupies the highest percentage in the Jackson Purchase, with substantial proportions in both the Pennyroyal and the West-

* There is a discrepancy of 40 acres between inventoried and noninventoried land as reported by the Soil and Water Conservation Committee.

A tobacco field in the Inner Bluegrass. The soil in this region, derived from limestone, is suitable for farming, especially the production of burley tobacco and rotational pasture.

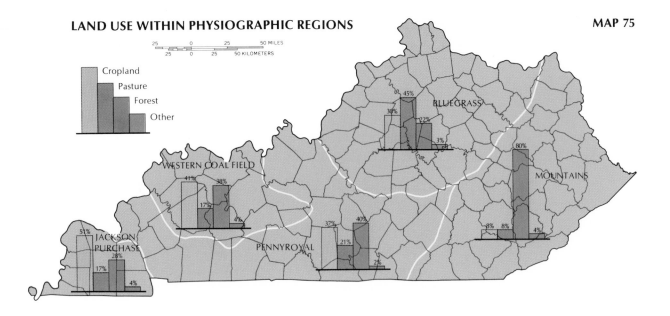

LAND USE WITHIN PHYSIOGRAPHIC REGIONS

MAP 75

25   0   25   50 MILES
25   0   25   50 KILOMETERS

Cropland
Pasture
Forest
Other

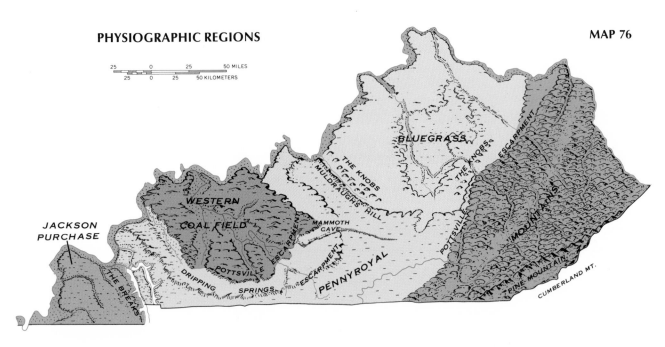

PHYSIOGRAPHIC REGIONS

MAP 76

25   0   25   50 MILES
25   0   25   50 KILOMETERS

110

ern Coal Field. The eastern Mountains have the least amount of land in crops.

Soils and topography influence the land use pattern in the state. Almost all soils in Kentucky, with the exception of stream deposits, have developed under forest cover and under essentially the same climate. Differences between soils are due chiefly to differences in parent materials, to topography, and to the length of time these materials have been exposed to weathering and other soil-forming processes. The various combinations of parent material, topography, and time of exposure are readily expressed by dividing the state into twelve areas (**Map 77**). The suitability of these land areas for various uses is shown on **Map 78.** The areas are described below.

Area 1: The *Purchase–Mississippi River Flood Plain* west of Hickman and in Fulton County is level and generally poorly drained. Soils have developed from Mississippi River sediments and are high in fertility. About 85 percent of the area is in crops.

Area 2: The *Purchase–Thick Loess Belt* west of the Tennessee River is undulating to rolling with about 25 percent bottomland and the remainder well-drained uplands. Soils are derived mainly from thick loess which lies over sand, gravel, and, in a few places, clay. Most soils are medium in fertility and are suitable for farming, but erosion is a problem even on gentle slopes. About 50 percent of the area is in cropland, 20 percent in pasture, 15 percent in woodland, and 15 percent idle.

Area 3: The *Cumberland–Tennessee Rivers Section* is a hilly section extending from about five miles east of the Cumberland River to about five miles west of the Tennessee River. Soils have generally developed in thin loess which lies over gravel and chert beds. Soils generally are low in fertility, droughty, and inferior for agriculture. Much of the area is suitable for and is used for recreational purposes.

Area 4: The *Western Coal Fields–Low Hills and Valley Areas* border the Ohio River from east of Owensboro to west of Sturgis, and extend to south of Providence, Madisonville, and Hartford. Upland soils have generally developed in loess over sandstones and shales. They are medium in fertility, suitable for farming, and resemble the upland soils of the Purchase. Bottomland soils are extensive and generally need drainage. Overflow soils of the Ohio River are generally fertile, dark silt loam and silty clay loams. They are neutral in reaction and well supplied with phosphate. About 65 percent of the area is in cropland, 25 percent in pasture, and 10 percent in woodland.

PHYSIOGRAPHIC AND MAJOR SOIL ASSOCIATION AREAS

MAP 77

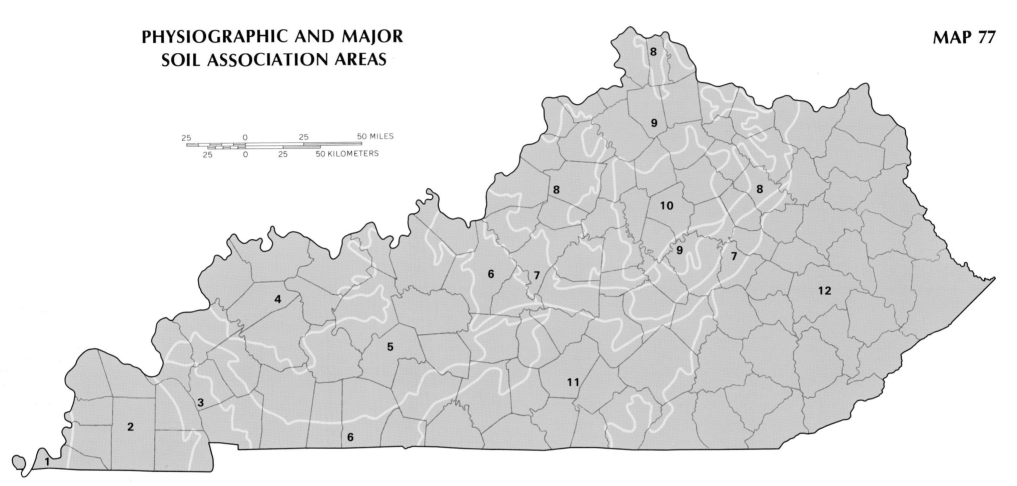

AREA	IMPORTANT SOIL SERIES	AREA	IMPORTANT SOIL SERIES
1. Purchase–Mississippi River Flood Plain	Commerce–Sharkey–Robinsonville	7. Knobs	Carmon–Colyer–Captina
2. Purchase–Thick Loess Belt	Grenada–Calloway–Falaya	8. Outer Bluegrass	Lowell–Shelbyville–Fairmount
3. Cumberland–Tennessee Rivers Section	Brandon–Lax–Guin	9. Hills of the Bluegrass	Eden–Faywood–Nicholson
4. Western Coal Fields–Low Hills & Valley Areas	Loring–Memphis–Falaya	10. Inner Bluegrass	Maury–Lowell–AcAfee
5. Western Coal Fields–Hilly Uplands, and the Sandstone–Shale–Limestone Area of the Western Pennyrile	Zanesville–Gilpin–Weikert–Caneyville	11. Eastern Pennyrile	Waynesboro–Baxter–Garmon–Bedford
6. Western Pennyrile–Limestone Area	Pembroke–Cumberland–Crider	12. Mountains and Eastern Coalfields	Shelocta–Jefferson–Rarden–Weikert

Area 5: The *Western Coal Fields–Hilly Uplands*, and the *Sandstone–Shale–Limestone Area* lie mostly south of Area 4 and extend in an arc from near Smithland through Princeton, Russellville, and Cave City to the Ohio River west of Brandenburg. Soils have generally developed from sandstone, shales, and loess, with some limestone in the southern part of the area, and are medium to low in fertility. Much of the topography is hilly, but numerous areas of undulating to level upland and considerable bottomland are present. The more level areas of upland and parts of the bottomland are poorly drained. About 30 percent of the area is in cropland, 25 percent in pasture, and 45 percent in woodland.

Area 6: The *Western Pennyroyal–Limestone Area* lies in a horseshoe-shaped belt to the south and east of Area 5. It is a gently sloping karst (sinkhole) area with soil developed from limestone of varying degrees of purity. Soils are generally well drained and well suited to general farming, but on the more irregular karst slopes erosion control is difficult. About 75 percent of the area is in cropland, 15 percent in pasture, and 10 percent in woodland.

Area 7: The *Knobs* separate the Pennyroyal and Mountains from the Bluegrass and extend from near Louisville southeast to Lebanon, east to Halls Gap in Lincoln County, and northeast to Bath County. The area is characterized by conical hills which are erosional remnants of surrounding uplands. The edge next to the Pennyroyal areas is hilly and rough, but toward the Bluegrass there are wide valley floors and bottomlands between the Knobs. Soils of the Knobs themselves are shallow with many rock outcrops and are poorly suited to farming. The valley soils are generally acid and low in organic matter, nitrogen, and phosphorus; many are poorly drained. When properly treated these soils produce fairly well. About 30 percent of the area is in cropland, 20 percent in pasture, and 50 percent in woodland.

Area 8: The *Outer Bluegrass* occurs in three discontinuous arcs around the outer edge of the Bluegrass Region. It is more rolling than the Inner Bluegrass and the soils generally contain less phosphate. The Outer Bluegrass is suited to general farming and in particular to pasture and hay crops. Good burley tobacco is produced on most of the soils. About 60 percent of the area is in cropland, 35 percent in pasture, and 5 percent in woodland.

Area 9: The *Hills of the Bluegrass* lie in a circle separating the Inner Bluegrass from the Outer Bluegrass and constitute a well-dissected plateau characterized by narrow winding ridges and valleys. Hillsides generally slope 20 to 30 percent and surface rock is common. Soils are rather high in phosphorus and high in lime and

LAND USE SUITABILITY

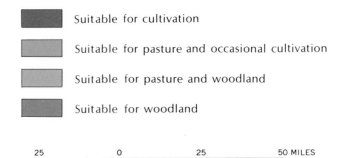

- ■ Suitable for cultivation
- ■ Suitable for pasture and occasional cultivation
- ■ Suitable for pasture and woodland
- ■ Suitable for woodland

25   0   25   50 MILES

25   0   25   50 KILOMETERS

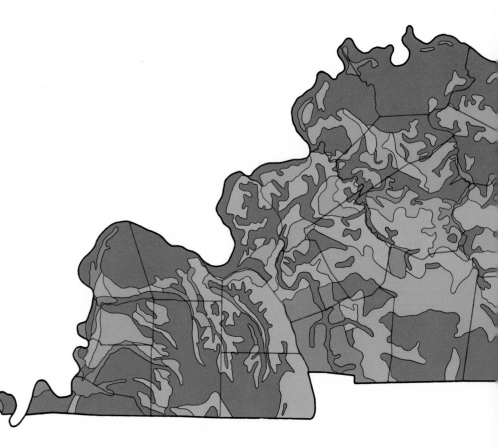

MAP 78

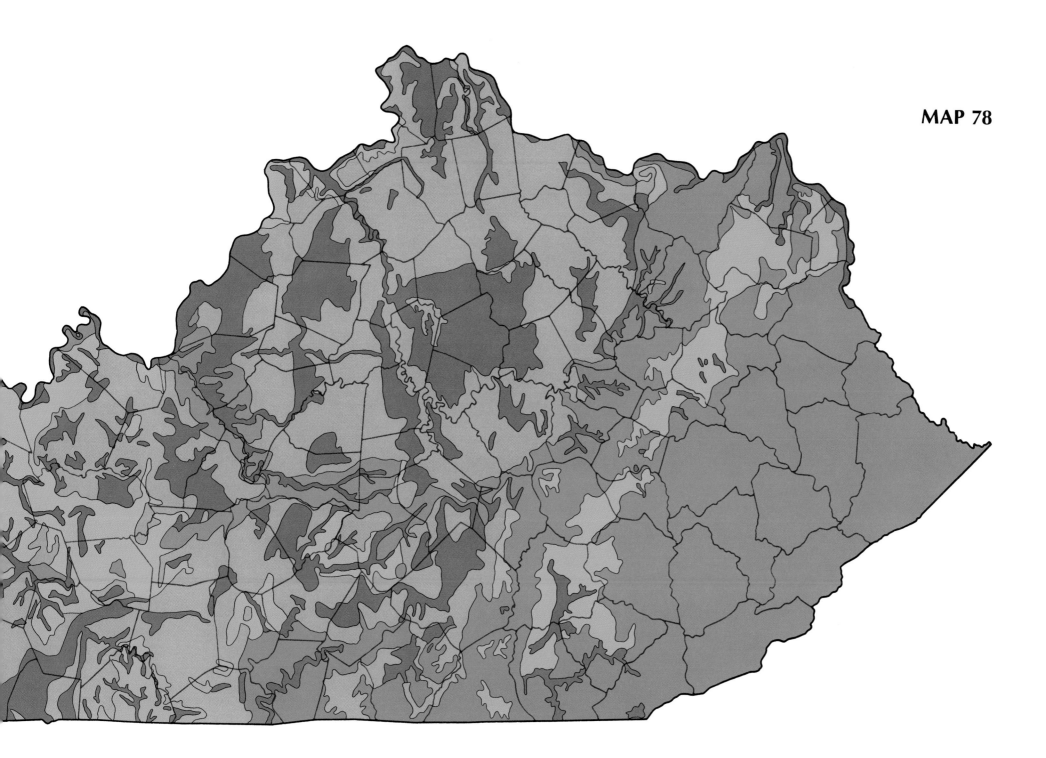

Above left: The canyon of the Russell Fork near the Kentucky-Virginia line. Thin rocky soils and rough topography make the Mountains poor for farming. The dominant land use is forest.

Above right: The Knobs surrounding the Outer Bluegrass, some of which rise as high as 900 feet. Soils on the hills, largely black or blue-gray shales, are shallow, stony, and poor. Valley soils are often compact and surface drainage is poor, but when properly managed, as in the foreground, they make good cropland.

Land Use & Physical
114 *Character*

potassium but are droughty. About 15 percent of the area is used for row crops, 75 percent for pasture, and 10 percent for woodland.

Area 10: The *Inner Bluegrass* comprises a circle in the middle of the Bluegrass and centers near Lexington. Most of the topography is gently rolling. Soils are derived mainly from phosphatic limestones and are suited to general farming and well suited to burley tobacco, alfalfa, hay crops, and pasture. On steeper slopes rock is generally close to the surface. About 80 percent of the area is used for cropland and rotational pasture, 18 percent for permanent pasture, and 2 percent for woodland.

Area 11: The *Eastern Pennyroyal* extends from the Knobs south of the Bluegrass southwest to the Tennessee border. The eastern portion is mostly a limestone-capped karst (sinkhole) plateau; elsewhere it is largely shale with associated limestones. Topography is undulating to hilly, with the Cumberland River area being rugged.

Soils are medium to low in fertility. About 30 percent of the area is used for row crops, 30 percent for improved pasture, and the remainder for woodland and recreational areas.

Area 12: The *Mountains and Eastern Coal Fields* lying east of the Bluegrass and Eastern Pennyroyal regions have three general areas: the more mountainous eastern area, the west-central plateau area, and an escarpment area on the western edge. Topography ranges from hilly to mountainous. Soils are generally low to very low in fertility. The plateau area has less rugged topography and some areas are rather well suited to general farming after lime and fertilizer are applied. The bottomlands are locally important to agriculture. About 7 percent of the entire area is used for crops, 12 percent for pasture, and 81 percent is in forest.

WILFORD A. BLADEN and HARRY H. BAILEY

MAP 79

RAINFALL VARIABILITY

Isolines in percent of rainfall variability,
plus or minus mean rainfall

Period of record 1951-1960

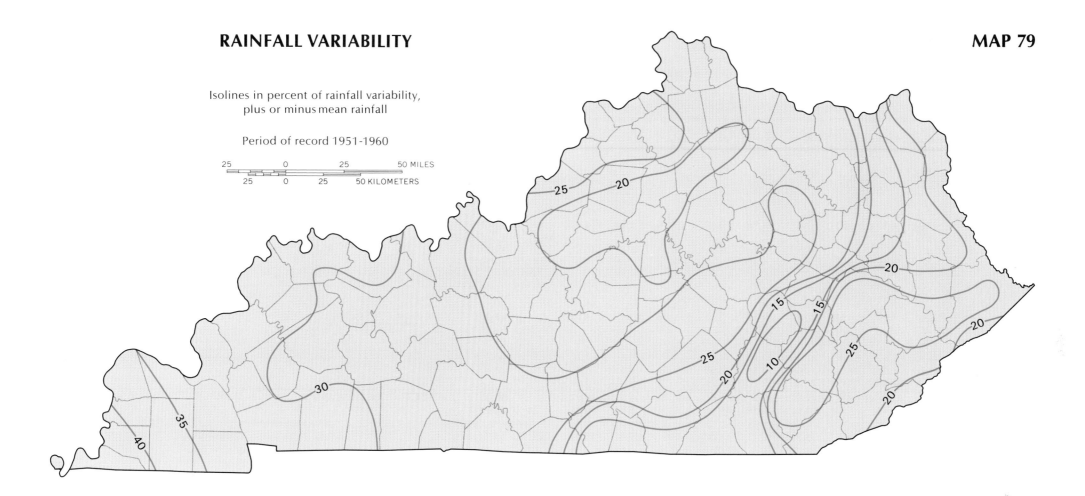

CLIMATIC CHARACTERISTICS

MAPS 79-88

Most elements of Kentucky's climate vary appreciably across the state. This is not surprising since Kentucky is more than 420 miles in length and spans over half the distance from the Mississippi to the Atlantic. Moreover, the variations in elevation influence the climatic characteristics. The state's geographical position is within the belt of westerly winds which bring a succession of low-pressure storm systems across the state. Because all low-pressure systems have winds which circulate counterclockwise, storms approaching the state from the west are accompanied by southwest winds bearing warm moist air from the Deep South states or the Gulf of Mexico. The mild but rainy low-pressure systems are periodically displaced by high-pressure, fair-weather systems. All high-pressure systems circulate clockwise and consequently bring northwest winds which are cool and dry. The frequent passage of these two types of systems through the state makes for changeable weather during each season of the year. The mountains to the east tend to block out weaker storms from that direction, though hurricanes in the Gulf of Mexico or the Atlantic often cause rain in Kentucky.

Rainfall reliability or variability **(Map 79)** is an important consideration for the farmer. Ironically, the most stable rainfall pattern in the state is in the region with the least agricultural potential. Rainfall is most regular or consistent along the Appalachian front, where a narrow row of counties enjoys less than 10 percent variation in annual rainfall. Rain varies the most in the Jackson Pur-

TABLE 8
Climatic Statistics
Average temperature in degrees Fahrenheit; precipitation in inches
Period of record: 1931-1974

	Jan.	Feb.	Mar.	Apr.	May	June	July	Aug.	Sep.	Oct.	Nov.	Dec.	Annual	Average annual snowfall (in inches)	Average date, last spring freeze	Average date, first fall freeze	Given number of days, last spring to first fall freeze
Paducah																	
Temp.	36.0	38.9	47.1	59.2	67.7	76.1	79.2	77.8	70.8	60.2	47.6	38.2	58.2	7.9	April 4	October 28	206
Prec.	4.05	3.49	4.87	4.12	4.88	4.03	3.66	3.27	3.28	2.63	3.78	3.63	45.69				
Murray																	
Temp.	36.8	39.8	47.8	59.5	67.6	75.7	78.5	77.7	70.7	60.2	48.1	39.3	58.5	7.0	April 4	October 28	205
Prec.	4.25	4.23	5.27	4.27	4.25	3.77	3.90	3.27	3.28	2.89	4.12	3.99	47.49				
Owensboro																	
Temp.	34.9	37.9	46.0	57.9	66.8	75.3	78.2	76.8	70.2	59.5	46.6	37.3	57.3	12.6	April 10	October 21	193
Prec.	4.01	3.51	5.06	4.33	4.38	3.80	3.45	3.08	3.24	2.30	3.59	3.53	44.28				
Bowling Green																	
Temp.	35.6	38.4	46.3	57.8	66.7	75.0	78.1	76.9	70.3	59.1	46.4	37.7	57.4	8.6	April 13	October 24	193
Prec.	4.62	4.42	5.46	4.18	4.17	4.24	4.12	3.01	3.02	2.39	4.02	4.44	48.09				
Mammoth Cave																	
Temp.	35.1	37.8	46.0	57.8	65.3	72.7	75.6	74.6	68.7	58.4	46.1	37.5	56.3	11.0	April 29	October 13	165
Prec.	4.90	4.40	5.45	4.28	4.17	4.68	4.19	4.02	3.39	2.55	4.22	4.35	50.60				
Louisville																	
Temp.	33.3	35.8	44.1	56.1	64.9	72.9	75.8	74.6	68.2	57.3	44.8	35.6	55.3	18.1	April 21	October 19	179
Prec.	3.53	3.47	5.05	4.10	4.20	4.05	3.76	2.99	2.94	2.35	3.33	3.34	43.11				
Lexington																	
Temp.	32.9	35.3	43.6	55.3	64.7	73.0	76.2	75.0	68.6	57.8	44.6	35.5	55.2	15.9	April 23	October 26	181
Prec.	3.95	3.42	4.80	3.87	4.16	4.31	4.83	3.40	2.65	2.12	3.36	3.62	44.49				
Covington																	
Temp.	31.1	33.3	41.7	53.9	63.2	72.1	75.6	74.4	67.8	56.8	43.8	33.7	54.0	23.6	April 19	October 22	185
Prec.	3.34	3.04	4.09	3.64	3.74	3.81	4.12	2.62	2.55	2.15	3.08	2.86	39.04				
Pikeville																	
Temp.	38.1	40.2	47.7	58.8	67.1	74.2	77.1	76.3	70.4	59.8	47.7	39.2	58.1	15.6	April 17	October 26	191
Prec.	3.38	3.62	4.32	3.60	3.84	4.12	5.06	3.50	3.41	2.15	2.91	3.30	43.21				
Ashland																	
Temp.	34.5	36.5	44.5	55.9	64.7	72.6	75.8	74.4	68.5	57.7	45.7	36.6	55.6	13.7	April 23	October 22	181
Prec.	3.24	2.90	4.04	3.40	3.98	3.75	4.16	3.50	2.89	2.08	2.90	2.99	39.83				
Maysville																	
Temp.	32.4	34.2	42.3	53.8	62.9	71.5	74.8	73.4	67.0	56.4	44.4	34.6	54.0	14.1	April 25	October 24	181
Prec.	3.56	3.20	4.62	3.81	4.03	3.92	4.57	3.75	2.97	2.25	3.33	3.22	43.23				
Berea																	
Temp.	36.1	38.3	46.3	57.7	65.9	73.4	76.2	75.4	69.4	59.0	46.8	38.1	56.9	14.7	April 17	October 20	185
Prec.	3.98	3.55	4.58	4.31	3.81	4.58	5.16	3.99	3.06	2.01	3.45	3.63	46.11				
Somerset																	
Temp.	36.4	39.4	46.2	56.5	65.0	72.3	75.3	74.4	68.5	57.5	46.3	37.5	56.3	14.7	April 23	October 17	176
Prec.	4.88	4.75	5.07	4.16	3.97	4.73	4.63	3.70	3.25	2.29	3.87	3.99	49.29				
Williamsburg																	
Temp.	37.8	39.8	47.3	57.9	66.1	73.2	76.0	75.0	69.2	58.6	47.2	39.4	57.3	8.8	April 21	October 18	174
Prec.	4.36	4.28	4.79	4.01	3.95	4.14	5.09	3.62	2.89	2.59	3.76	4.05	47.53				

chase, where the difference in the total amount of rainfall from one year to the next may be 40 percent. Prolonged drought is rare, and fortunately for the farmer the annual variations tend to repeat for several consecutive years.

Annual precipitation for the state averages about 45 inches but varies from about 40 inches in the north at Covington to about 50 inches in a zone from Mammoth Cave to Somerset in the south-central portion of the state **(Map 80)**. Areas having more than 50 inches of precipitation per year are shaded on Map 80. Rain falls year-round **(Table 8)**, although the fall months tend to be dry, with September and October averaging 20 to 35 percent less rainfall than the heavy rainfall months of January and March. A dry fall season is advantageous to farmers who wish to harvest corn or soybeans or cut and house tobacco. The heavy winter rains present a possible erosion hazard on clean-tilled fields, such as corn or tobacco. To prevent the winter loss of topsoil many farmers plant winter wheat on the open fields in the fall; the wheat roots, although dormant, prevent soil from washing downslope. The northern and eastern parts of the state have the highest number of days annually with measurable precipitation **(Map 81)**. Areas with more than 140 days of measurable precipitation per year are shaded on the map.

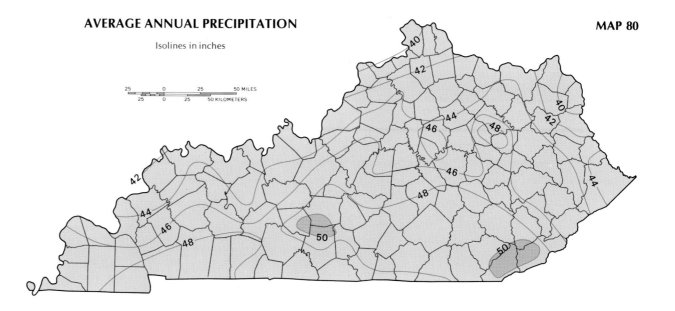

AVERAGE ANNUAL PRECIPITATION

Isolines in inches

MAP 80

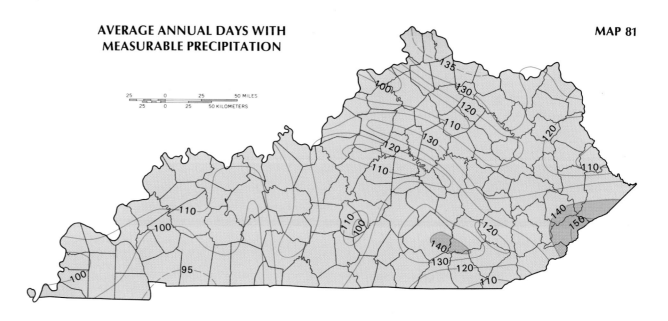

AVERAGE ANNUAL DAYS WITH MEASURABLE PRECIPITATION

MAP 81

Fields flooded by the Kentucky River in Estill County. Spring and summer rains sometimes cause flash floods in relatively small areas. Winter floods are apt to be more widespread. Flood damages are increasing with more intensive use of floodplains.

Climate 117

JANUARY, MEAN MAXIMUM TEMPERATURE

Isolines in degrees fahrenheit

MAP 82

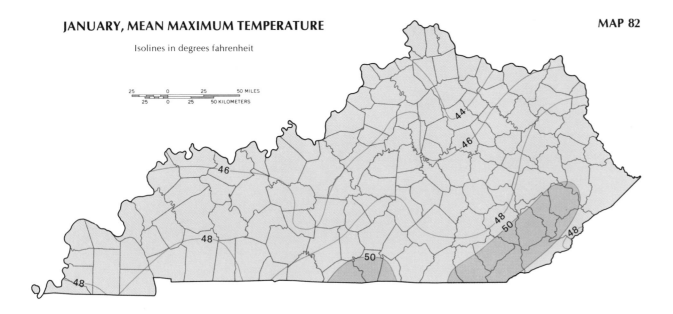

JANUARY, MEAN MINIMUM TEMPERATURE

Isolines in degrees fahrenheit

MAP 83

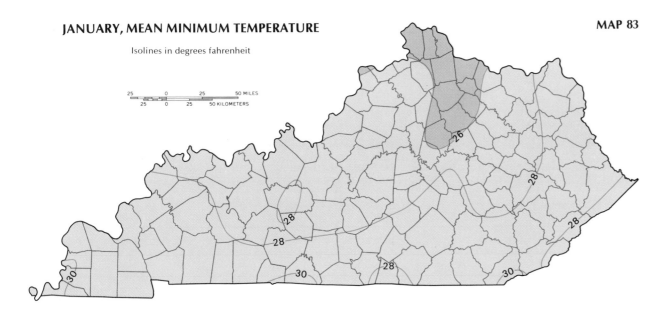

Kentucky's winters are usually comparatively mild. In general the coldest areas are northern Kentucky and higher elevations in the eastern mountains. Average snowfall varies from as high as twenty-four inches per year at Covington to as low as seven inches at Murray. The lowest temperature ever recorded in the state was −34.6° F, an extremely rare occurrence.

High-pressure systems bring the winter's coldest temperatures. Temperatures below 0° F are not uncommon but rarely last for more than a day or two. Because high-pressure systems circulate clockwise, the pattern of January temperatures is much different from the July pattern. In January, temperatures decrease from south to north **(Maps 82 and 83)**. In July, temperatures decrease from west to east **(Maps 84 and 85)**. In winter the state's warmest stations are at Pikeville, Williamsburg, and Murray. The coolest is at Covington along the Ohio border. Covington also gets almost two feet of snow each winter, more than any other station in the state. Areas with the highest January temperature and with the lowest mean minimum temperature have been shaded on Maps 82 and 83.

In summer the maximum temperature can reach 100° F or more across the state, but the mean maximum does not exceed 92° F in any part of the Commonwealth. The coolest section of the state in summer, in terms of both mean maximum and mean minimum temperatures, is the eastern Mountains. There the altitude of 2,000 feet or more causes the temperatures to be as much as six degrees cooler than those in the Purchase. On Map 84 the area with the highest July mean maximum temperature is shaded; likewise the area with the lowest mean minimum temperature is shaded on Map 85.

JULY, MEAN MAXIMUM TEMPERATURE

Isolines in degrees fahrenheit

MAP 84

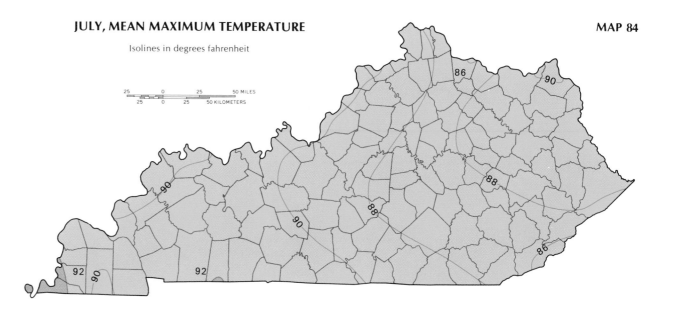

JULY, MEAN MINIMUM TEMPERATURE

Isolines in degrees fahrenheit

MAP 85

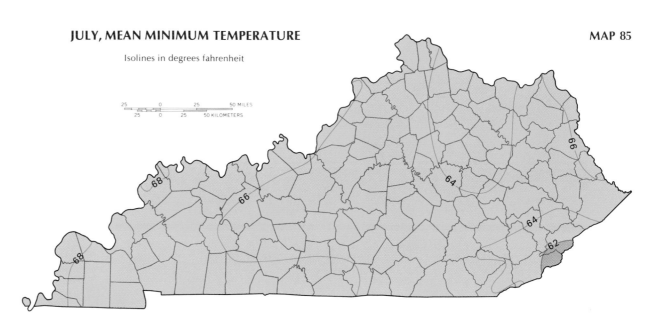

July and August are usually the warmest months in Kentucky, with temperatures occasionally reaching 100° F at most locations. "The sun shines bright on my old Kentucky home" is not an exaggerated statement in view of the large amount of sunshine in the state. The warm weather favors recreational activities on rivers and lakes, such as boating on Greenbo Lake (left).

AVERAGE DATE, LAST SPRING FREEZE (32°f)

MAP 86

Period of record 1931–1960

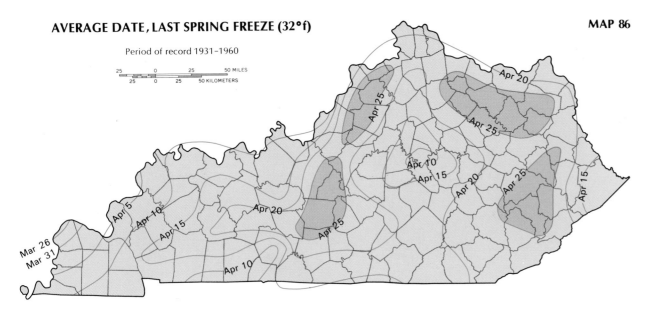

AVERAGE DATE, FIRST FALL FREEZE (32°f)

MAP 87

Period of record 1931–1960

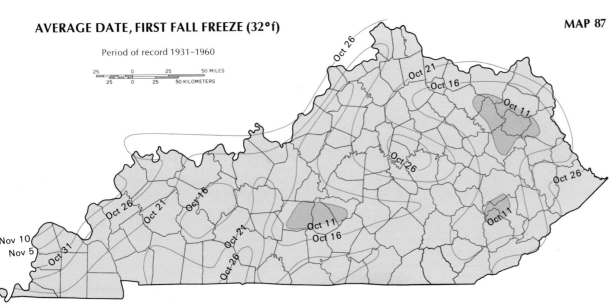

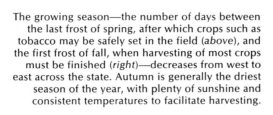

The growing season—the number of days between the last frost of spring, after which crops such as tobacco may be safely set in the field (*above*), and the first frost of fall, when harvesting of most crops must be finished (*right*)—decreases from west to east across the state. Autumn is generally the driest season of the year, with plenty of sunshine and consistent temperatures to facilitate harvesting.

Land Use & Physical Character

Isolines in total degrees annually (sum of departures of
daily average temperature below 65°F) (18.3°C)
Period of record 1931–1960

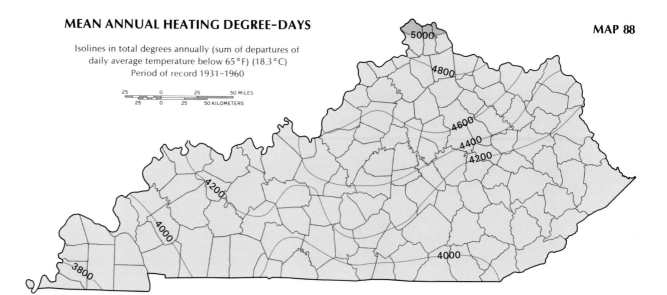

The growing season, or the period between the last spring freeze **(Map 86)** and the first freeze of fall **(Map 87)**, varies over 25 percent across the state. Areas with the last spring freeze late are shaded on Map 86; similarly, areas experiencing the first fall freeze very early are shaded on Map 87. The average length of the frost-free season ranges from over 200 days in the Jackson Purchase, the warmest section of the state, to 164 days in the eastern Mountains. The cooler Mountains are not only higher in altitude than the far west, but lie at a greater distance from the active low-pressure system storm track which brings warm moist air from the Gulf. In addition to enjoying a longer frost-free period, the western and southern counties also enjoy a warmer growing season, as indicated by fewer heating degree-days **(Map 88)**. Hickman in Fulton County, for example, has a total heat resource for crops which is about 21 percent greater than that at Burlington in Boone County. This helps explain how cotton, a crop with an extended growing season, can be grown in the Jackson Purchase, and how soybeans, a fast maturing crop, can be double-cropped with winter wheat, which is harvested in late June.

KARL B. RAITZ and WILFORD A. BLADEN

Above left: Ice storms in Kentucky can occur any time between November and March but the ice seldom remains more than a few days.

Left: Damage from one of the severe tornadoes that struck central Kentucky in April 1974. Several years may pass without a severe tornado or several may visit the state in a single year. Kentucky averages one damaging tornado per year.

XII. FORESTRY AND AGRICULTURE

MAP 89

FORESTRY

Kentucky is a transition zone between northern and southern forests. Southern red oak, shortleaf pine, pecan, bald cypress, and many other species representative of the southern forest, as well as sweet birch, American basswood, swamp white oak, and white pine of the northern species, are found here. The blending of northern and southern species makes Kentucky's forest flora rich with a wide variety and number of tree species.

The first white men found Kentucky largely timbered although some extensive grassland areas, such as the Green River Barrens, did exist. Woodlands varied greatly from dense forests in the eastern highlands to the more open forests of the Bluegrass area, where cane, wild rye, and clover formed a dense herbaceous ground cover. The geographical differences in the original forests of Kentucky are well documented by E. Lucy Braun; her intensive field study of remnant virgin forests in the 1920s and 1930s still provides the best scientific information available on this subject.

After nearly 200 years of settlement and use, about 11 million acres, or 43 percent of the state, are still in woodland **(Map 89)**. As would be expected, more forest is found in the rugged portions of the state. In eastern Kentucky, most counties have 60 to 85 percent of the land in trees. Except in the Bluegrass, forests still represent a significant land use throughout the state.

A small decrease in forest acreage has probably occurred in recent years. Land has been converted from forest for new highways, power lines, reservoirs, subdivisions, and other construction. Many bottomland forests along the Green and other western Kentucky rivers have been cleared for soybean production in response to high agricultural prices. Strip mine activity has claimed many thousands of acres throughout the major coal fields. It is doubtful if abandonment of marginal agricultural land and reclamation by natural reforestation or tree planting has kept pace with conversion to other uses.

Kentucky's forests in 1975 contained some 26 billion board feet of sawtimber. Various species of oak make up about half of the total. The wood-based industries contribute heavily to the economy of the Commonwealth. They are widely distributed throughout the state and consist of a wide variety of types of businesses: a considerable number of hardwood dimension plants, furniture manufacturers, and other wood industries.

These industries had a combined capital investment of more than $277 million in 1973. They furnished direct employment to some 18,000 people in 1975, with an annual payroll of more than $106 million; this did not include the many people employed in harvesting and handling timber products. In 1973 annual sales for the wood-based industries topped the half-billion-dollar mark for the first time in history.

JAMES A. NEWMAN and WILFORD A. BLADEN

LAND IN COMMERCIAL FOREST

MAP 89

Over 80 60-79 40-59 20-39 Under 20

Interval in percent of total land area

25   0   25   50 MILES
25   0   25   50 KILOMETERS

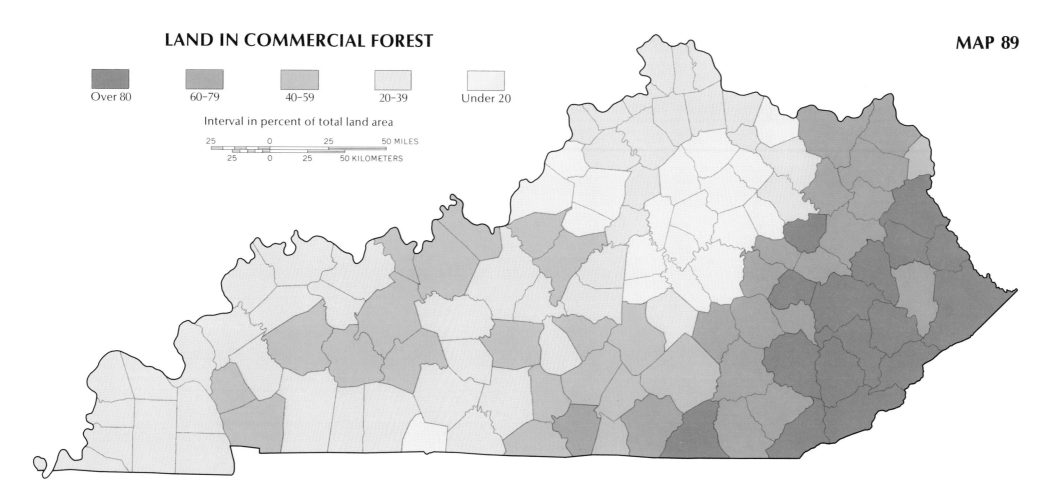

Most of Kentucky's commercial forest is in the Mountains. Forest industries such as lumbering (*left*) employ about 18,000 people with an annual payroll of more than $100 million. Modern methods are used to cut and transport commercial timber to sawmills, such as the automated one at Morehead (*far left*). Thousands of Kentucky farmers receive income from forest products. Conservation and efficient management of forest resources can contribute to economic prosperity and employment stability in the state.

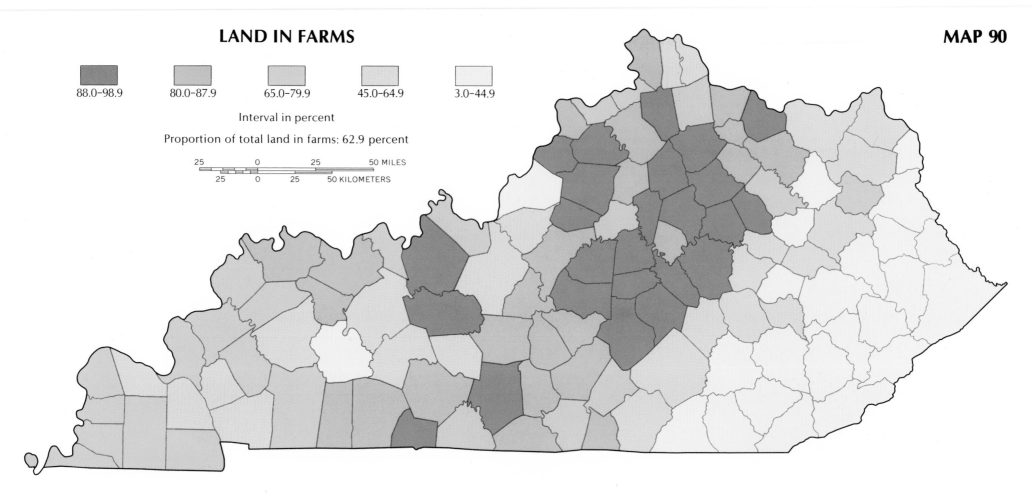

MAP 90

LAND IN FARMS

88.0-98.9   80.0-87.9   65.0-79.9   45.0-64.9   3.0-44.9

Interval in percent

Proportion of total land in farms: 62.9 percent

25   0   25   50 MILES

25   0   25   50 KILOMETERS

MAPS 90-93   FARMLAND AND FARM TYPES

Between 1970 and 1975 the importance of agriculture in Kentucky's economy declined in terms of employment, while manufacturing, government, trade, and services markedly increased. Among these major categories of economic activity, agriculture was the lowest employer, with 88,218 persons in 1974 out of a total work force of 1,347,000. Throughout large sections of the state, however, farming is the predominant form of land use. The regions with the most land in farms are the Inner and Outer Bluegrass counties, which generally have 88 percent or more of the land in farms (**Map 90**). Much of the eastern and central Pennyroyal also has a high proportion of land in farms. The Mountains, with steep slopes and stony soils, have the lowest percentage of land in farms. While two-thirds to three-fourths of the Purchase land is in farms,

more could be added by reclamation of marshy areas. Altogether, some 16,200,000 acres in Kentucky are in farmland.

The Purchase, the central Pennyroyal, and five counties in the Western Coal Field are important areas of harvested cropland (**Map 91**). Between 17.6 and 45 percent of the land in these counties is in forage, feed grain, or other commercial crops. Much of the farmland in the Bluegrass and eastern Pennyroyal, where only about 15 percent of the land is cultivated, is in permanent pasture. Second- or third-growth woodland predominates in eastern Kentucky, where only a tiny fraction of the land is tilled. Harvested acreage of principal crops in 1974 totaled 4,633,000 acres, a 9 percent increase over 1973 acreage. Wheat registered the largest acreage increase.

TOTAL LAND AREA IN HARVESTED CROPLAND

MAP 91

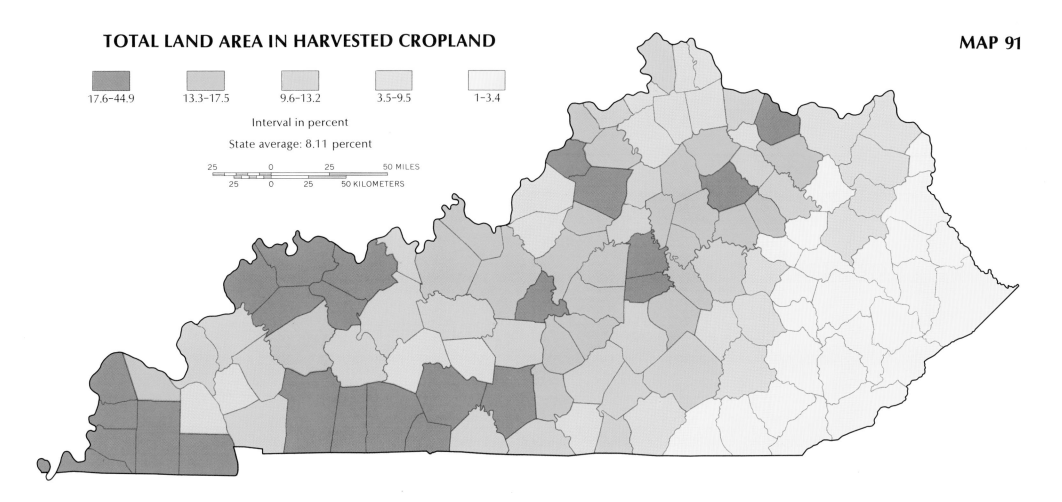

17.6-44.9   13.3-17.5   9.6-13.2   3.5-9.5   1-3.4

Interval in percent

State average: 8.11 percent

25   0   25   50 MILES
25   0   25   50 KILOMETERS

Far left: A valley bottom farm in Rowan County. Because of steep topography and low soil fertility, the total area in farms in the Mountains is low. Rural residents in eastern Kentucky are usually semisubsistence farmers who have supplementary employment in coal mining or forest industries.

Left: A farm in Barren County in the Pennyroyal. The topography consists of gently rolling land, used mostly for farming.

Farmland & Farm Types 125

MAP 92

FARMLAND IN CLASS 1-5 FARMS

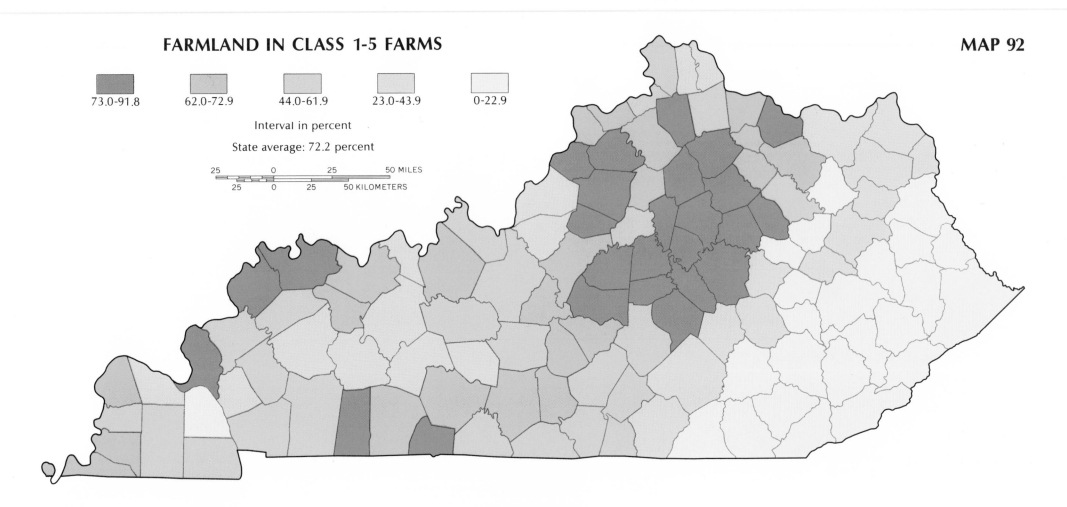

73.0-91.8   62.0-72.9   44.0-61.9   23.0-43.9   0-22.9

Interval in percent

State average: 72.2 percent

25   0   25   50 MILES

25   0   25   50 KILOMETERS

A farm in the Knobs, a transitional area between the Mountains and the Bluegrass. Farms in this agricultural area dotted with small communities are substantially larger than Mountain farms to the east, but the region has fewer farms in classes 1 to 5 than does the Bluegrass, the Pennyroyal, or the Jackson Purchase.

Kentucky had 126,000 operating farms in 1974 and, along with Tennessee and Illinois, was ranked fifth nationally in number of farms, behind Texas, Missouri, Iowa, and North Carolina. Farms in classes 1 to 5 **(Map 92)** are "real" farms in that they produce at least $2,500 worth of commodities annually. Only 48 percent of Kentucky's farms are in these classes, yet they produce 90 percent of all farm products sold. In the Bluegrass, in the counties around Henderson and Owensboro, and in the central Pennyroyal, from 62 to 92 percent of the farmland is in classes 1 to 5 farms. The Mountain counties have the smallest percentage of acreage in these classes of farms.

On **Map 93,** counties are classified according to their predominant source of income from farming. If a farmer sells tobacco, for example, and his income from that crop amounts to 50 percent or more of all his income from farm products during the year, his

CENSUS FARM TYPES

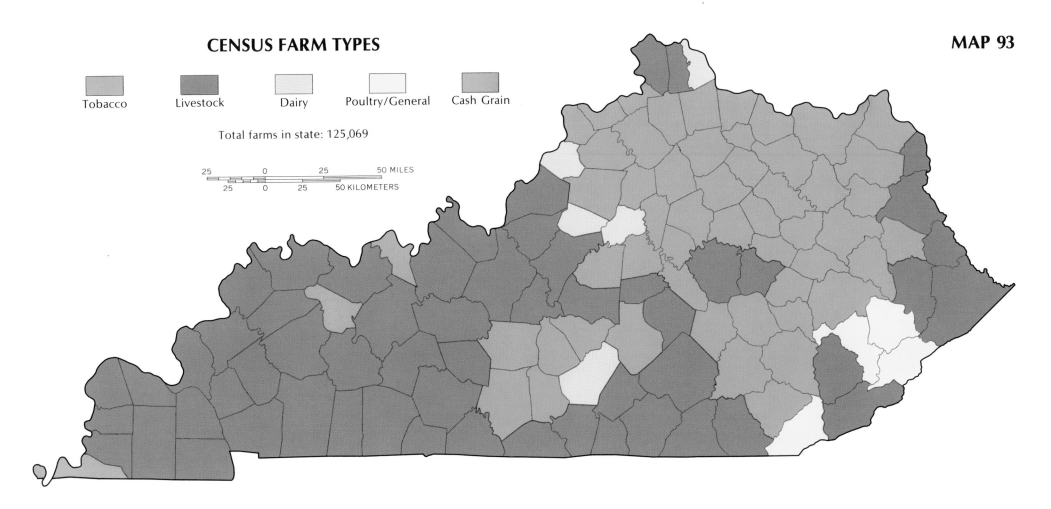

Tobacco Livestock Dairy Poultry/General Cash Grain

Total farms in state: 125,069

25 0 25 50 MILES
25 0 25 50 KILOMETERS

MAP 93

farm is classified as a tobacco farm. Although tobacco is the most important cash crop in Kentucky, farmers dependent on it as their primary source of income are concentrated in the central and eastern parts of the state. Livestock farms producing beef cattle, hogs, and sheep predominate in the Pennyroyal, the Western Coal Field, and the Purchase. Specialization in dairy products is concentrated on modern farms near large urban centers or lingers as a holdover from general farming in marginal farming areas. Three decades ago many Kentucky farms were general farms. They were small in size and produced a wide variety of grains, hay or silage, hogs, and poultry, and maintained a dairy cow or two. Today small farms cannot compete with the larger, more specialized units, and general farming predominates only in Perry and Bell counties.

KARL B. RAITZ and WILLIAM A. WITHINGTON

A barn on a tobacco farm near Burkesville in Cumberland County. Kentucky tobacco farms produced over 431 million pounds of burley in 1975. For almost a century Kentucky has maintained its position as the foremost producer of burley, which accounted for 47.6 percent of the total value of crop production in 1975.

Farmland & Farm Types 127

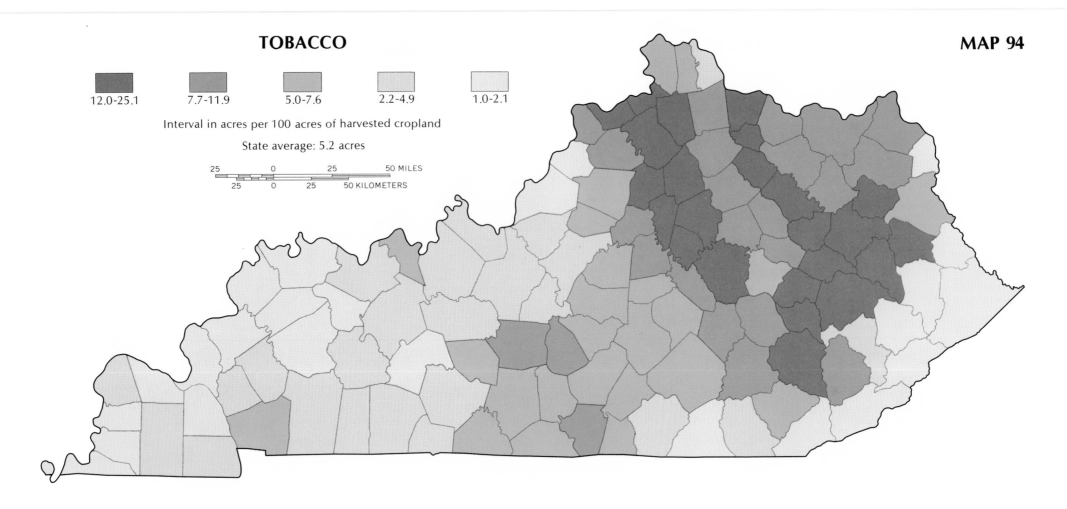

12.0-25.1   7.7-11.9   5.0-7.6   2.2-4.9   1.0-2.1

Interval in acres per 100 acres of harvested cropland

State average: 5.2 acres

25   0   25   50 MILES

25   0   25   50 KILOMETERS

MAPS 94-99   MAJOR CROPS AND PASTURELAND

Tobacco (**Map 94**) is Kentucky's primary cash crop. Tobacco acreage in 1974 was 189,050, or 4 percent of the total harvested cropland. Of this acreage, 93 percent was in burley tobacco. The value of all tobacco production was $485 million, 27 percent of the market value of all agricultural products sold. Production of this labor-intensive crop requires about 25 percent of the state's farm labor, or about 10 million man-hours per year. About 20 percent of all tobacco grown in the United States, and nearly 70 percent of all burley tobacco, is grown in Kentucky. In terms of crop value Kentucky ranks second in tobacco in the United States.

Because of its high value per acre, tobacco is especially important to Kentucky's small and medium-size farms. The production of burley today is centered in the Bluegrass, with lesser amounts in the adjoining counties of the Mountains, where the relatively small amount of harvested cropland is dominated by small valley tobacco fields. As **Map 94** indicates, tobacco, being labor-intensive, does not compete well with crops grown on the mechanized cash grain farms of the western Pennyroyal. Small amounts of dark air-cured tobacco, a chewing or cigar-filler variety (USDA type 35), are grown in Christian, Todd, and Logan counties. Green River, also a dark air-cured variety used in chewing tobacco and snuff (USDA type 36), is produced in counties near Henderson and Owensboro. A third dark tobacco is produced in the Jackson Purchase. It is fire cured in air-tight barns to give it a distinctive color and aroma.

CORN

MAP 95

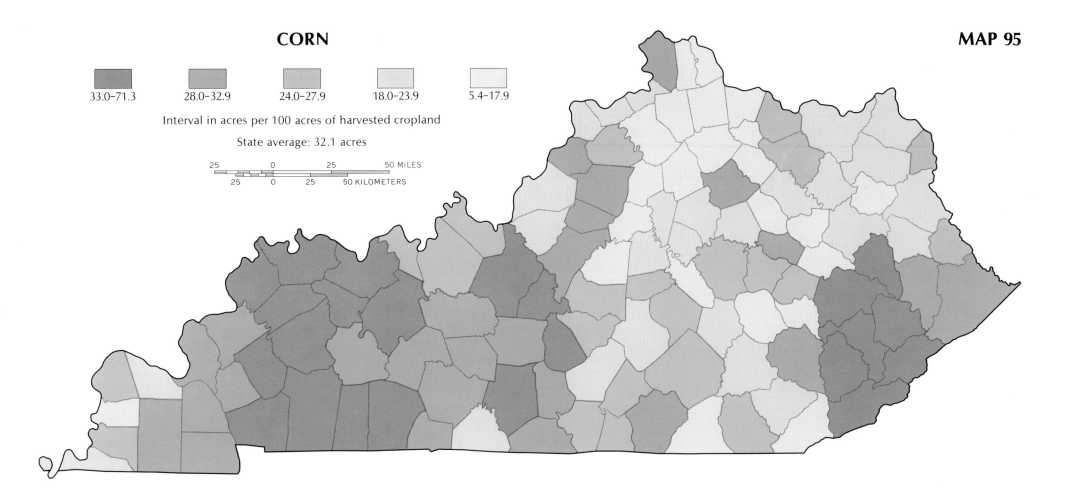

33.0–71.3 28.0–32.9 24.0–27.9 18.0–23.9 5.4–17.9

Interval in acres per 100 acres of harvested cropland

State average: 32.1 acres

25   0   25   50 MILES

25   0   25   50 KILOMETERS

Although most of Kentucky lies south of the American Corn Belt, corn, both for grain and for silage, was the state's second crop in both acreage and value of production **(Map 95)**. In 1974 corn occupied 1,260,000 acres of harvested cropland, nearly 27 percent of the total cropland. It had a value of $333 million, 18.5 percent of the value of all agricultural products sold. In acreage corn was exceeded only by hay, and in value, only by tobacco.

The leading corn-producing area, with over 33 percent of the harvested cropland, lies in the Western Coal Field and the western Pennyroyal. In these areas corn production is associated with the raising of feeder pigs and hogs **(Map 103)**. Although corn occupies a high percentage of cropland in many Mountain counties, production is small since cropland is limited. In the Bluegrass the yield of corn per acre is generally above the state average but far more land is devoted to pasture and tobacco than to corn.

Every county in Kentucky has some acreage in corn but the major concentrations are in the Western Coal Field and the western Pennyroyal. In 1975 corn occupied 1.3 million acres and accounted for 21.7 percent of the total value of crop production in the state.

Major Crops & Pastureland   129

SOYBEANS

MAP 96

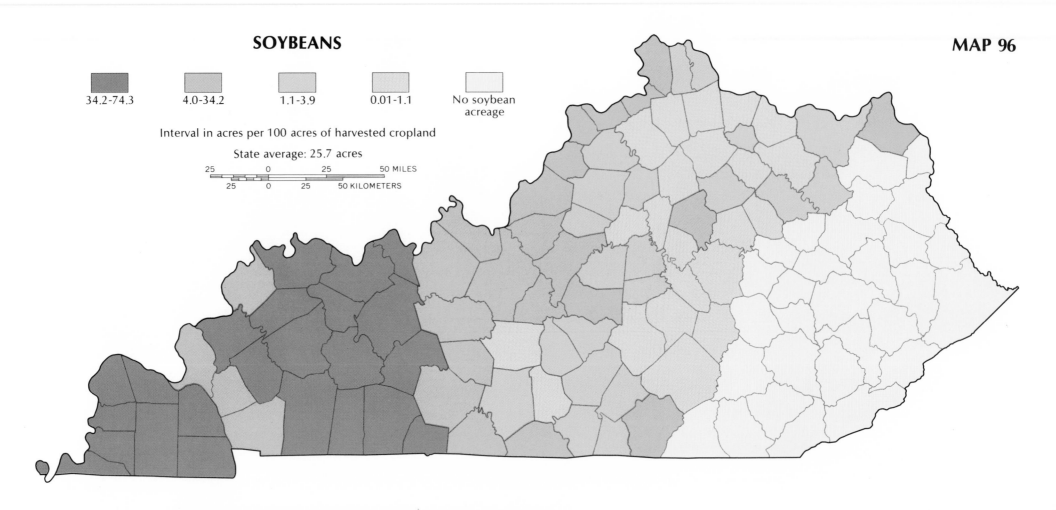

34.2-74.3	4.0-34.2	1.1-3.9	0.01-1.1	No soybean acreage

Interval in acres per 100 acres of harvested cropland

State average: 25.7 acres

25   0   25   50 MILES
25   0   25   50 KILOMETERS

Soybeans, rapidly gaining in importance, rank as one of Kentucky's major crops, accounting for 14.5 percent of the total value of crop production in 1975. They occupied 1.2 million acres, yielding 31.8 million bushels. Soybeans do well on the rich alluvial bottomlands of the Jackson Purchase and Western Coal Field.

Soybeans now account for more than a quarter of Kentucky's agricultural land in major harvested crops **(Map 96)**. Of the 1974 total of 4,662,650 acres harvested in major crops, soybeans accounted for 1,170,000 acres, yielding 29,250,000 bushels. Soybean production had a value of $227 million and accounted for 12.6 percent of the value of all agricultural products sold. The recent rapid and large-scale expansion of harvested soybean acreage in Kentucky has resulted in an eastward diffusion of soybean production from the Jackson Purchase, which a few years ago had the only sizable soybean acreages in the state, into the Western Coal Field and much of the Pennyroyal. If soybeans continue to be a highly profitable cash crop their production may increase in other areas of Kentucky.

HAY

MAP 97

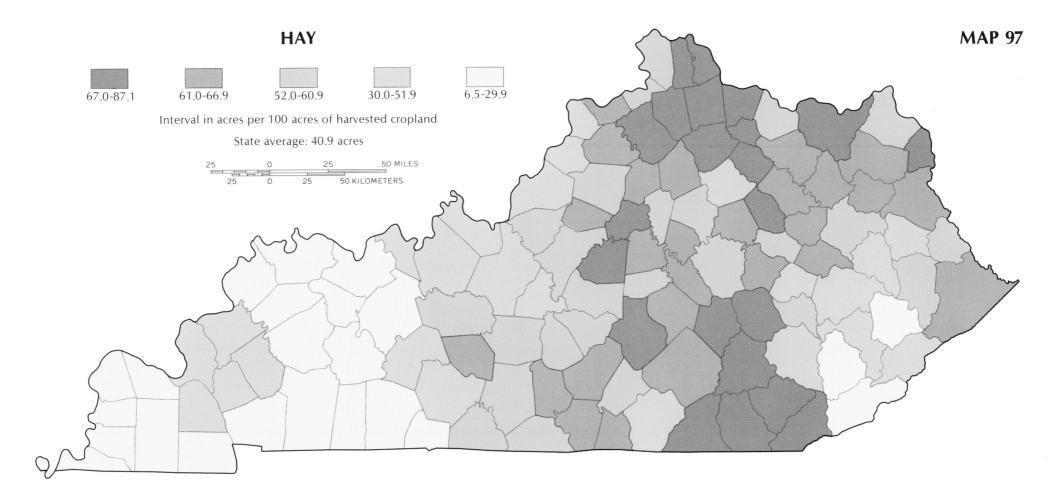

67.0-87.1 61.0-66.9 52.0-60.9 30.0-51.9 6.5-29.9

Interval in acres per 100 acres of harvested cropland

State average: 40.9 acres

25   0   25   50 MILES
25   0   25   50 KILOMETERS

Hay crops come from about 32 percent of the state's harvested cropland **(Map 97)**. Alfalfa and other hay occupied 1,497,000 acres in 1974, the largest acreage of any crop in the state. All hay crops for feed were valued at $114 million, or about 6.3 percent of the value of all agricultural products sold. As a percentage of harvested cropland, hay is generally more important in east-central Kentucky than in the western portion. Traditionally hay is cut and stored in barns for winter feed, but the current trend is to bale it in large rolls and leave it in the field for livestock to feed on.

Large acreages of hay are also harvested for seed, which is used to raise hay for feed. These acreages are not included in the above figures. In 1974 they were devoted mainly to tall fescue (36,000 acres), lespedeza (10,000 acres), and red clover (6,100 acres). The total value of hay grown for seed in 1974 was $2,371,000.

Hay is one of Kentucky's major crops. In 1975 it accounted for 10.9 percent of the total value of crop production and occupied 1.5 million acres. Hay is produced in every county but is most important in the Bluegrass. Alfalfa is one of the principal hay crops raised to feed cattle, both dairy and beef.

WHEAT

MAP 98

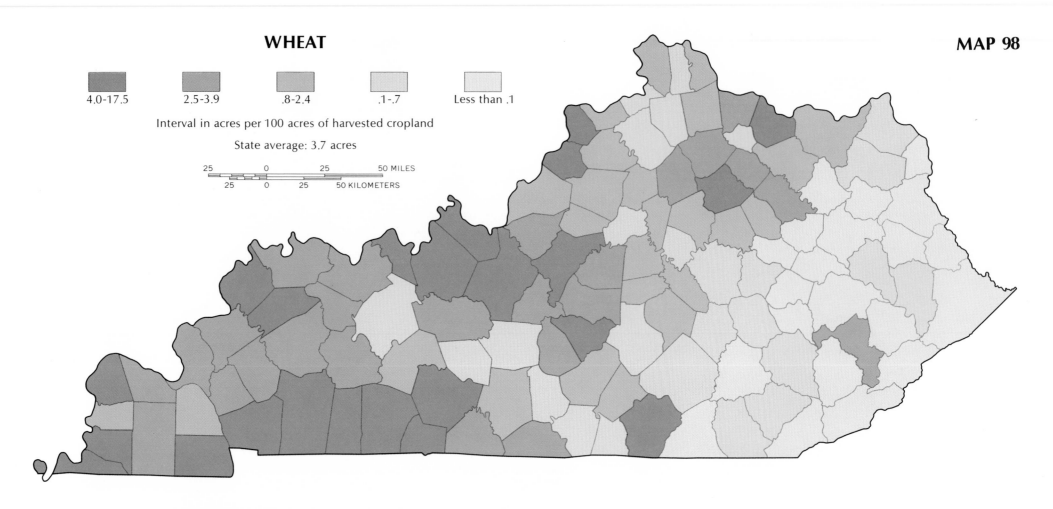

4.0-17.5	2.5-3.9	.8-2.4	.1-.7	Less than .1

Interval in acres per 100 acres of harvested cropland

State average: 3.7 acres

25   0   25   50 MILES
25   0   25   50 KILOMETERS

A farmer plowing the land for wheat. In Kentucky wheat is sown in the fall as a winter ground cover. It may be plowed under to help enrich the soil in spring or it may be harvested in early summer. Kentucky farmers harvested 12 million bushels of wheat from 352,000 acres in 1975. In most counties wheat uses only a small portion of the harvested cropland.

Wheat in 1974 was harvested on 390,000 acres in Kentucky, representing 8.4 percent of total harvested cropland **(Map 98)**. Wheat production had a value of $46.6 million, making it the fifth leading crop in the state. Wheat accounts for almost 10 percent of the state's harvested cropland and about 2.6 percent of the total value of agricultural products. More than half of Kentucky's harvested acreage in wheat is in the west-central part of the state. The Purchase is the second largest wheat-producing area. Most of Kentucky's wheat is of the soft red winter varieties, with lesser acreages devoted to hard red winter wheat. A variety known as Abe is by far the highest yielding variety of those grown in Kentucky. Other small grains grown in the state account for less than 1 percent of the value of agricultural products sold.

PASTURELAND

MAP 99

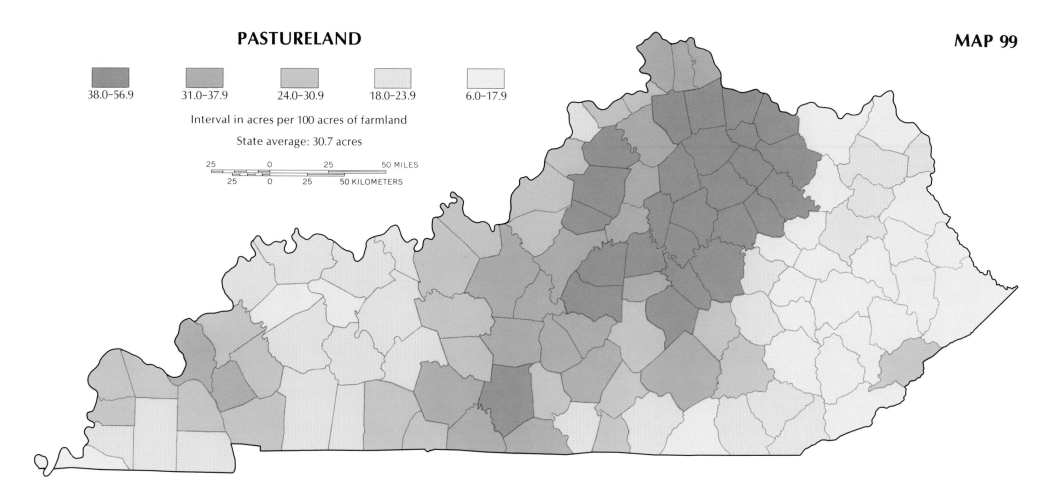

38.0–56.9 31.0–37.9 24.0–30.9 18.0–23.9 6.0–17.9

Interval in acres per 100 acres of farmland

State average: 30.7 acres

25 0 25 50 MILES

25 0 25 50 KILOMETERS

Pastureland in Kentucky is second in acreage only to forests. It is concentrated primarily in the Bluegrass and the south-central Pennyroyal **(Map 99)**. Most pastureland in the Bluegrass is devoted to horse farms. Often purebred cattle are also raised on the horse farms for pastureland control. Most of the Mountain counties have less than 18 percent of the farmland in pasture, and farmland represents a small percentage of the total area of the region. Many Mountain farmers raise small numbers of livestock for family use. Farmland in western Kentucky is mostly in crops rather than pasture. Kentucky's nickname comes from the dominance of bluegrass (*Poa pratensis*) in the state's pastures.

WILLIAM A. WITHINGTON and KARL B. RAITZ

Dairy cows on pastureland, those farm fields where grass is allowed to grow for grazing. Pastured lands in Kentucky occupy more acres than crops. While animals may graze on many kinds of coarse grasses and weeds, most pasturelands are planted to special kinds of grass, such as bluegrass and fescue, which are hardy and nutritious.

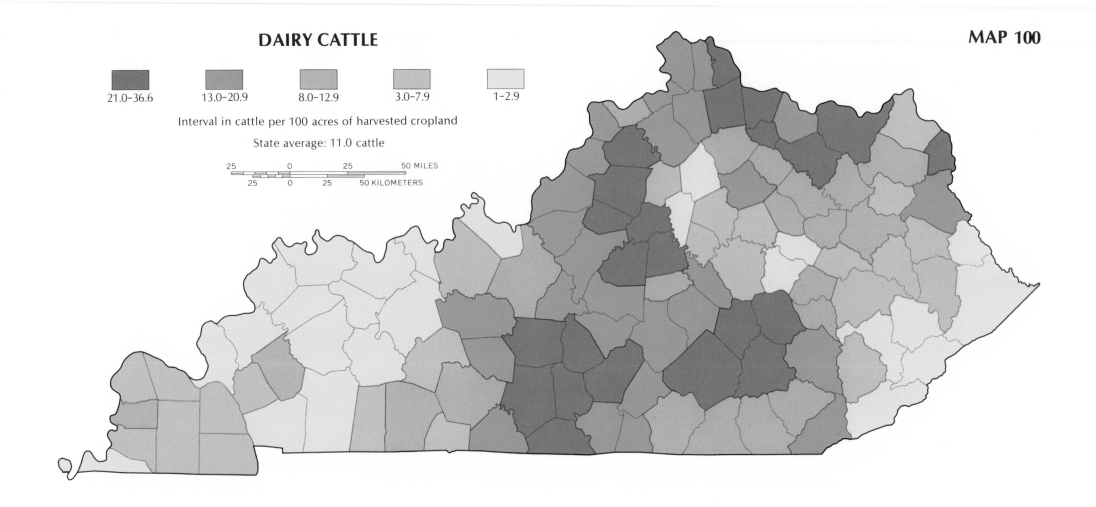

DAIRY CATTLE

21.0–36.6   13.0–20.9   8.0–12.9   3.0–7.9   1–2.9

Interval in cattle per 100 acres of harvested cropland

State average: 11.0 cattle

25   0   25   50 MILES
25   0   25   50 KILOMETERS

MAP 100

MAPS 100-106   COMMERCIAL LIVESTOCK

Dairy cattle in Kentucky are most important in the eastern Penny-royal and the Outer Bluegrass (**Map 100**). Dairy products represent about 20 percent of the state's total receipts from livestock products. Although the numbers of dairy cattle have declined in recent years, the value of dairy products has increased. In 1974 Kentucky ranked eleventh in the nation in the number of dairy cattle. The state average was 6.4 dairy cattle per 100 acres of harvested cropland, a decline from the 1969 figure shown on the map. The most popular breeds are Holstein, Guernsey, Jersey, and Ayreshire.

Much of the fluid milk is produced near large urban markets

(**Map 101**). Shelby, Spencer, and Nelson counties are the main suppliers of the Louisville market, while Pendleton and Bracken counties are the main Kentucky sources for the Cincinnati metropolitan area. Nashville gets part of its supply from the southern Pennyroyal. Lesser urban centers supplied by Kentucky producers are scattered throughout the state. The 1974 production of 2.4 billion pounds of milk and cream was worth $194 million, about 11 percent of the value of all agricultural products sold. Much of Kentucky's milk production goes into the manufacture of butter, cheese, ice cream, and evaporated milk.

134   *Forestry & Agriculture*

MAP 101

FLUID MILK SOURCE REGIONS

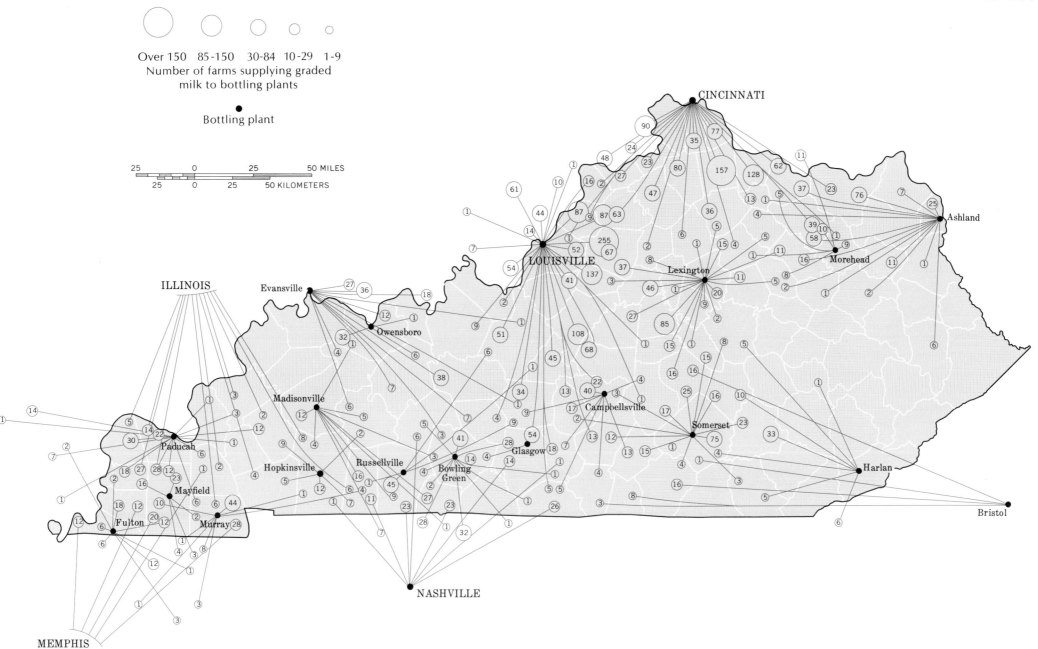

Over 150   85-150   30-84   10-29   1-9
Number of farms supplying graded
milk to bottling plants

● Bottling plant

25   0   25   50 MILES

25   0   25   50 KILOMETERS

CINCINNATI

90

77

35

48   24

23

80

157   128

62   11

16

27

13   1   5   37   23   76

7

25

47

11

Ashland

61

1

10

36

39   10

58   1   9

44

87   63

5

16

Morehead

7

14

255

2   1   8

1

52

8   1   15   4   11   5   2

67

1   2

ILLINOIS

1

54

137

37

Lexington

11   8

1   6

Evansville

27   36

18

41

46   1

20

9

3

12

1

85

Owensboro

32   1

4

2

51

108

1   15   15

1

9   6

68

16   15

16

38

45   1

25   16   10

34

40   3

1

Madisonville

6

13   22   4

17   16

5

12

17   Campbellsville   17

Somerset   23

33

8   4

1   41

28   54

13   12

75

9

1

Glasgow   18

13   15

1

Russellville

6   3   14

7

Bowling   14

Harlan

Hopkinsville

16   1

Green

1

4

14   1   14

12   45   9

27

4   1

5

1

8

3

1

Paducah

1

2

12

6

9   4

1   1   3

5

14

12

7   1   11

26

14   22

2

5

8

30

18   27   28   12

23

23   23

3

2   16

Mayfield

1

45

28

32

1

18   12   10

44

20

12   6   Fulton   12   Murray   28

2

6

7

8

4

12

1   1

NASHVILLE

3

MEMPHIS

3

Bristol

6

CATTLE AND CALVES

MAP 102

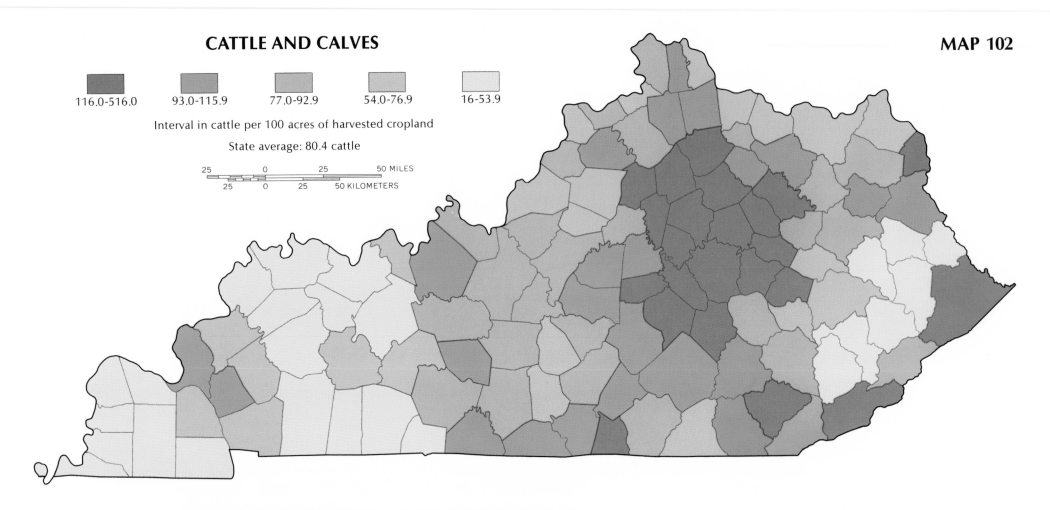

116.0-516.0   93.0-115.9   77.0-92.9   54.0-76.9   16-53.9

Interval in cattle per 100 acres of harvested cropland

State average: 80.4 cattle

25   0   25   50 MILES

25   0   25   50 KILOMETERS

The number of beef cattle and calves in Kentucky has been increasing each year. The value of those sold in 1975 was $313 million. During most of the year cattle can live off pastures, but for about three months in winter, when the weather is too cold for grass to grow as fast as the animals crop it off, they must be supplied with feed.

The raising of beef cattle is becoming increasingly important in Kentucky agriculture. The state ranked eleventh in the nation in 1974 in the number of cattle and calves. The most popular breeds are Hereford, Black Angus, Charolais, and Simmental. The largest concentration of cattle and calves in Kentucky (**Map 102**) is in the Bluegrass, where many of the cattle are raised as registered pure-bred animals for sale as breeding stock. Cattle and calves are important also in the eastern Pennyroyal but are relatively unimportant in the Mountains when one bears in mind that this region has little cropland. In 1974 the value of cattle and calves sold in Kentucky was $219 million, representing 12 percent of the value of all agricultural products sold.

HOGS

MAP 103

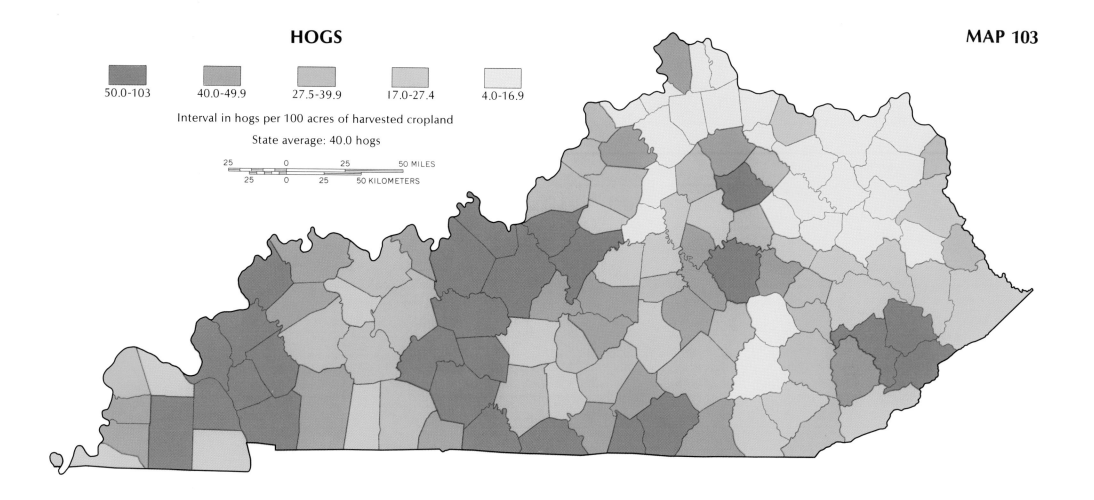

50.0-103 | 40.0-49.9 | 27.5-39.9 | 17.0-27.4 | 4.0-16.9

Interval in hogs per 100 acres of harvested cropland

State average: 40.0 hogs

In 1974 Kentucky farmers raised 1.1 million feeder pigs and hogs, approximately 24 per 100 acres of harvested cropland (**Map 103**). The state ranked thirteenth in the nation in the number of hogs. The value of hogs produced in 1974 was almost $163 million, or close to 9 percent of the value of all agricultural products sold. The number of hogs declined between 1969 and 1974 but increased in 1975 because of increasing demand and rising prices. Concentrations of hogs in Kentucky, both in absolute numbers and in number per 100 acres of harvested cropland, are mostly in two zones in the western half of the state. The most prolific breeds raised in Kentucky include Yorkshire, Duroc, American Landrace, Chester White Hampshire, Spotted Poland China, Berkshire, and Tamworth. Kentucky is famous for the production of country-cured ham, country sausage, and bacon.

Pigs and hogs in Kentucky are concentrated in the western part of the state, where desirable feeds such as corn are produced in quantity. With over one million hogs in 1976, Kentucky ranked thirteenth in the nation. Sows may have two litters per year, usually with six to twelve pigs to the litter. In six months these young pigs can be marketed profitably.

LIVESTOCK SALES OUTLETS

MAP 104

★ Federally inspected slaughter plants
◉ Custom slaughter plants (federally inspected)
⬕ Non-federally inspected slaughter plants
● Auction markets, weekly sales
▲ Stockyards, daily sales
■ Other stockyards

25   0   25   50 MILES

25   0   25   50 KILOMETERS

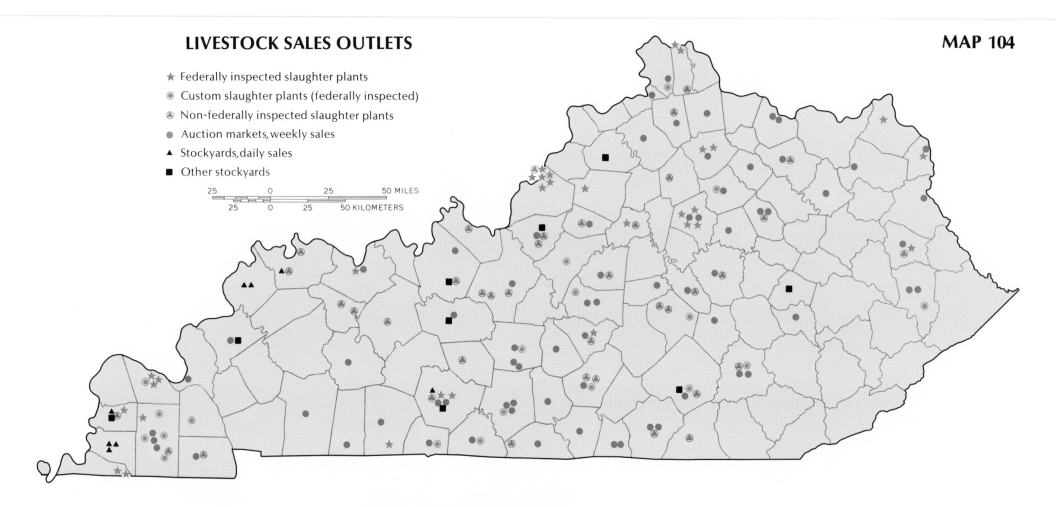

A cattle auction market in western Kentucky. Livestock is marketed at centers throughout the state except in the extreme southeastern portion. Cash receipts from the sale of livestock were $139 million in 1975. In that year Kentucky's 135 commercial plants slaughtered 5,300 calves, 394,700 cattle, and 1.45 million hogs.

Most federally inspected slaughter plants are located in or near larger urban areas **(Map 104)**. Livestock are trucked to slaughter plants for processing and distribution. Weekly auction markets are widely dispersed throughout most of the state, usually in the county seats and larger towns, but are fewer in the Mountains and in the extreme western Pennyroyal. Smaller stockyards primarily serve farmers in their local areas and are often associated with small slaughter plants. In 1974 commercial plants in Kentucky slaughtered 2,600 calves, 213,900 cattle, and 1,664,000 hogs. The total value of livestock production (cattle, calves, and hogs) in 1974 was about $382 million.

HORSES

MAP 105

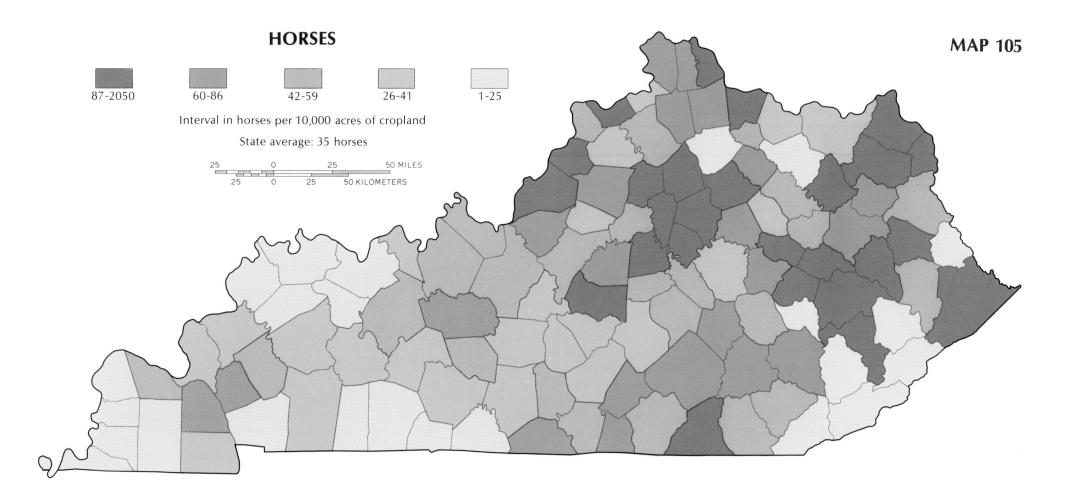

| 87-2050 | 60-86 | 42-59 | 26-41 | 1-25 |

Interval in horses per 10,000 acres of cropland

State average: 35 horses

25   0   25   50 MILES

25   0   25   50 KILOMETERS

Most of the horses of Kentucky are pleasure horses and are concentrated in the Inner Bluegrass, around Louisville, and in northern Kentucky near Covington (**Map 105**). The apparent high concentration of horses per 10,000 acres of cropland in the Mountains reflects the small amount of cropland in that region.

Pleasure horses raised in Kentucky in 1972 included Saddlebreds (21,000), Thoroughbreds (20,000), Tennessee Walking Horses (16,000), Quarter Horses (6,000), Standardbreds (3,000), and Appaloosas (1,600). Saddlebreds and Tennessee Walkers are used mainly for riding, Thoroughbreds and Standardbreds (trotters) for racing. Quarter Horse and Appaloosa racing is new to Kentucky.

A mare and her foal at Big Sink Farm in the Inner Bluegrass. The presence of great sires at stud on Bluegrass farms attracts the nation's best mares to Kentucky to be bred. Most Kentucky stallions stand on a live foal basis—the breeding fee is due when the foal stands and nurses. Fees range from $500 to $25,000. Most outstanding runners are syndicated for stud duty.

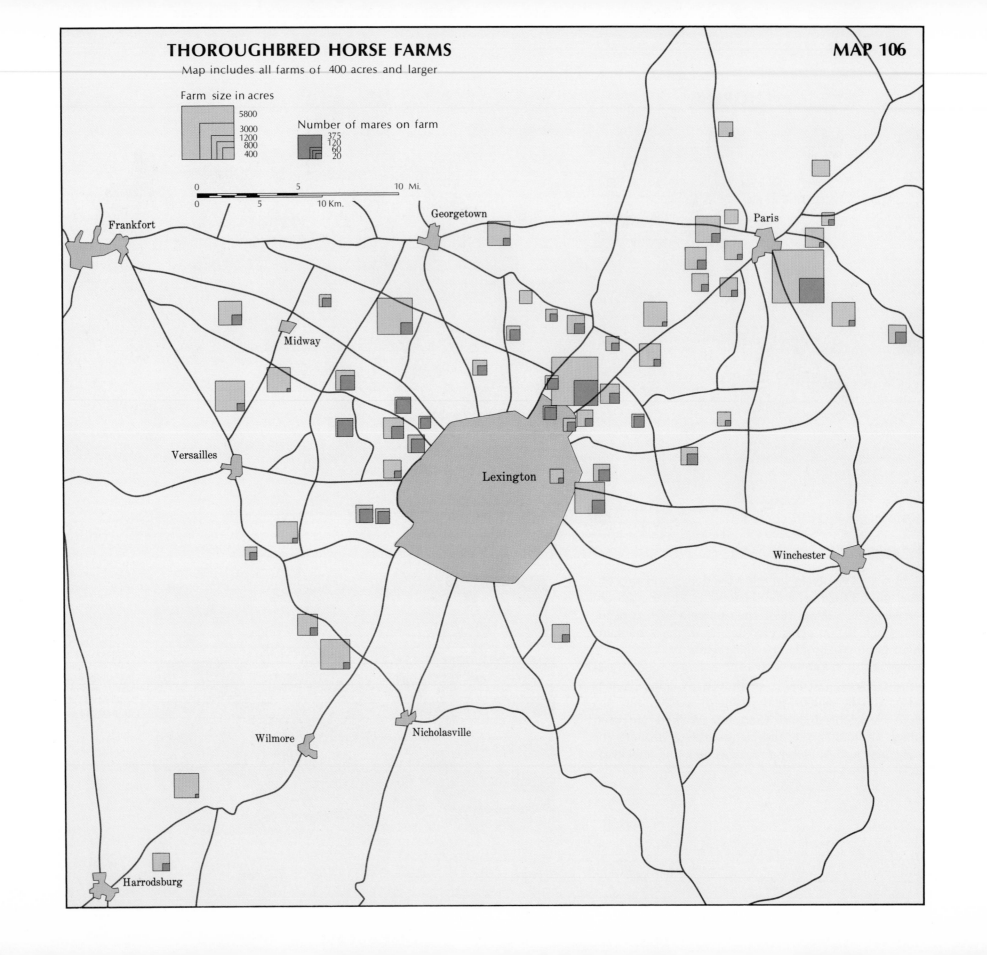

THOROUGHBRED HORSE FARMS

MAP 106

Map includes all farms of 400 acres and larger

Farm size in acres

5800
3000
1200
800
400

Number of mares on farm

375
120
60
20

0   5   10 Mi.

0   5   10 Km.

Frankfort

Georgetown

Paris

Midway

Versailles

Lexington

Winchester

Wilmore

Nicholasville

Harrodsburg

The most prominent and valuable horse raised in Kentucky, the Thoroughbred, has been bred primarily to race. In 1974 there were 351 Thoroughbred farms in Kentucky, with major concentrations near Lexington (**Map 106**). The number of Thoroughbreds in Kentucky has increased in recent years, from 15,693 in 1957 to 20,120 in 1972. Nearly half the Thoroughbred yearlings sold in the United States in 1972 were raised in Kentucky. The expansion of the Thoroughbred industry in other states, however, has reduced Kentucky's share of the national crop.

Breeding, racing, and sales of Thoroughbreds have a major impact on the economy of Kentucky. In 1973 the 83,707 acres devoted to Thoroughbreds accounted for $1.3 million in property taxes, state sales tax on Thoroughbred stud fees totalled $2.8 million, and pari-mutuel taxes on Thoroughbred racing amounted to $6.9 million. Boarding and training facilities generate additional tax revenues. Other specialized industries, such as equine insurance, equine education, veterinary care, trade journals, and tourism are significant factors in the economy of the Bluegrass Region. Excluding tourism, the investment in Kentucky racehorses and related industries is over one billion dollars, approximately 10 percent of the value of the entire American horse industry.

WILLIAM A. WITHINGTON and KARL B. RAITZ

Keeneland Sales in Lexington (*above left*) conducts four public Thoroughbred auctions annually. The July yearling sale and the November breeding stock sale are the most prestigious of their kind in the world. Individual horses may bring over a million dollars.

Above: Thoroughbreds being led past one of the barns on Calumet Farm near Lexington.

141

The value of products sold from this Pennyroyal farm with modern buildings and equipment is comparatively high. Preliminary cash receipts from farm marketings in Kentucky were $1.3 billion in 1975.

This small semisubsistence hill farm near Morehead has limited cropland and modest improvements. Its output is lower in market value than that from farms in other regions.

FARM INCOME, INVESTMENT, AND OWNERSHIP

In Kentucky the market value of all agricultural products sold has been increasing rapidly in recent years. In 1969 the state average per farm was $6,155 (**Map 107**). By 1974 the value per farm had risen to $11,907. This trend reflects in part the decline in the number of farms, down from 128,000 to 126,000 between 1969 and 1974, and the rise in the average size of farms, from 128 to 130 acres. Regions with the highest market value of all agricultural products sold are the Bluegrass, the western Pennyroyal, and the bottomlands along the Mississippi and Ohio rivers. The value of all produce per farm in these areas is several times higher than that in the eastern Pennyroyal and much higher than that in the Mountains.

In 1974, crops accounted for 68 percent of the total value of all agricultural production in the state ($1.8 billion), and livestock products accounted for the remainder. (See **Figure 2.**) Among the crops, tobacco, corn, soybeans, hay, and wheat were the most important; cattle and calves, milk and dairy products, and hogs led in the value of livestock production.

Figure 2. Value of Agricultural Production, 1974
(Value in millions of dollars)

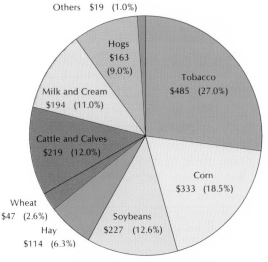

Others $19 (1.0%)
Hogs $163 (9.0%)
Milk and Cream $194 (11.0%)
Cattle and Calves $219 (12.0%)
Wheat $47 (2.6%)
Hay $114 (6.3%)
Soybeans $227 (12.6%)
Corn $333 (18.5%)
Tobacco $485 (27.0%)

Total value: $1,801,118,000

MARKET VALUE OF ALL AGRICULTURAL PRODUCTS SOLD

MAP 107

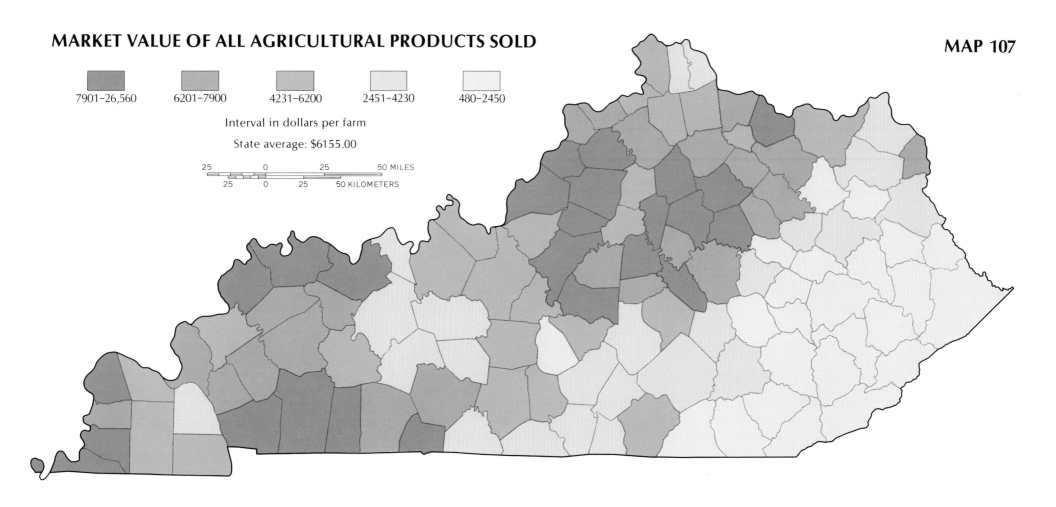

7901–26,560 6201–7900 4231–6200 2451–4230 480–2450

Interval in dollars per farm

State average: $6155.00

25   0   25   50 MILES

25   0   25   50 KILOMETERS

Far left: A tobacco auction in Lexington, the world's largest burley market. The state's 1975 burley production was 431 million pounds, valued at $460 million.

Left: Cattle on a farm at Eighty Eight. Livestock raising is an important element in the state's agricultural production. The value of cattle and calves in 1975 was $313 million.

Farm Income, Investment,
& Ownership   143

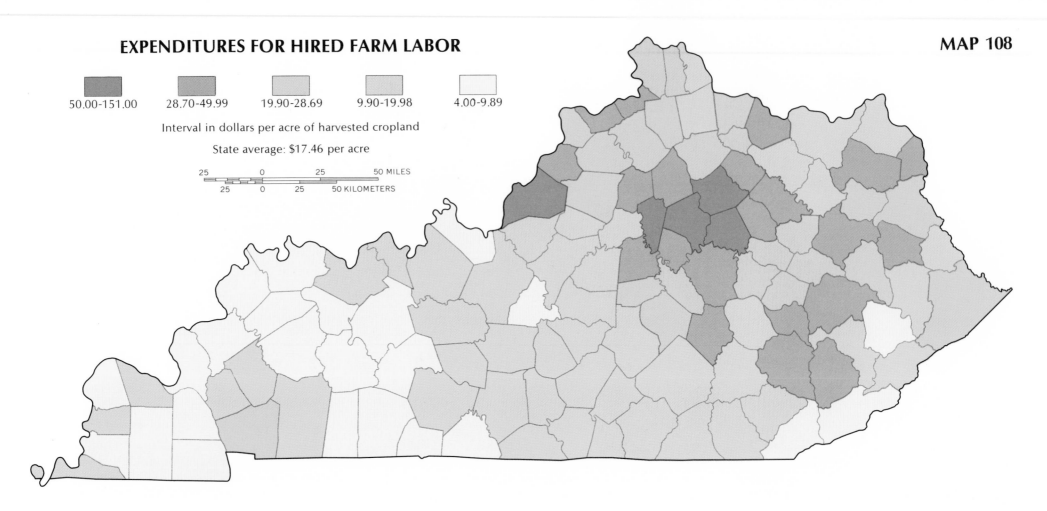

EXPENDITURES FOR HIRED FARM LABOR

MAP 108

50.00-151.00   28.70-49.99   19.90-28.69   9.90-19.98   4.00-9.89

Interval in dollars per acre of harvested cropland

State average: $17.46 per acre

25   0   25   50 MILES
25   0   25   50 KILOMETERS

Harvesting and care of tobacco require more labor than do other crops. Much farm labor is hired to perform the time-consuming processes of cutting and stripping tobacco before it can be air cured in tobacco barns. The average daily farm wage rate in Kentucky was $15.28 in 1975.

Farms with large tobacco acreages and others such as the horse and beef cattle farms of the Inner Bluegrass, spend several times the state average for hired farm labor (**Map 108**). On the large farms of western Kentucky, in contrast, mechanization has reduced the amount spent for labor. In 1973 expenditures for hired farm labor amounted to $16.51 per acre of harvested cropland for the state as a whole, a slight decrease from the 1969 figure of $17.46 per acre. Such expenditures represented about 12 percent of farm operating expenses. The average daily farm wage rate for 1974 was $13.50, about 10 percent higher than for 1973. While the average number of farm laborers in 1974 was 33,000, during the period of tobacco harvest the number rose to about 61,000, reflecting the labor-intensive character of the state's primary cash crop.

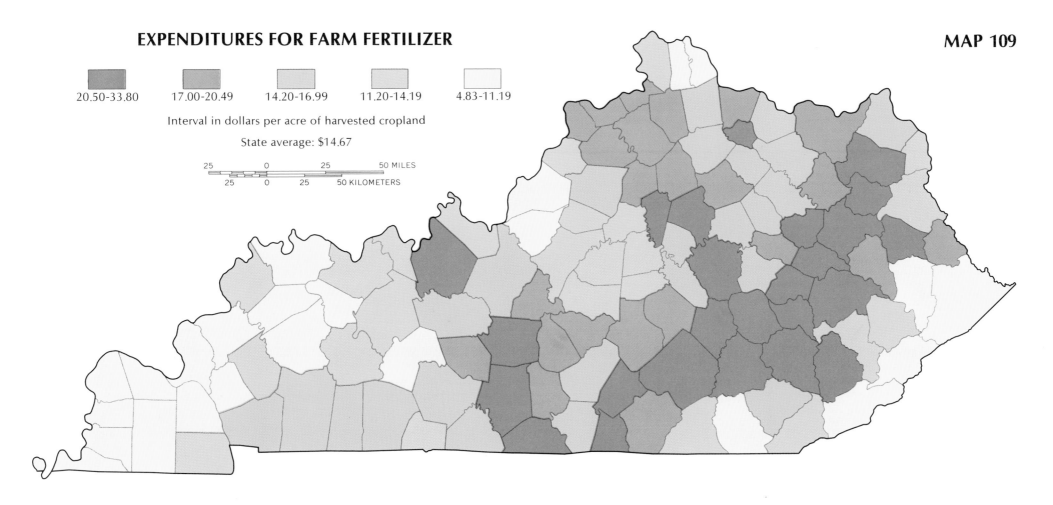

20.50-33.80	17.00-20.49	14.20-16.99	11.20-14.19	4.83-11.19

Interval in dollars per acre of harvested cropland

State average: $14.67

25   0   25   50 MILES

25   0   25   50 KILOMETERS

Expenditures for fertilizer, another major farm expense, amounted to $62.3 million in 1973, or nearly 10 percent of total farm expenses. Kentucky farmers' use of fertilizer increased from 521,000 tons in 1969 to 845,000 tons in 1973. Primary plant nutrients (nitrogen, phosphoric acid, and potassium oxide) totaled 300,081 tons in 1973. Approximately 75 percent of the fertilizer was applied during the spring months. The major use of fertilizer is for tobacco and corn, but the amount applied per acre is about four times as great for tobacco as for corn. The pattern of fertilizer expenditures **(Map 109)** reflects heavy applications in the central and eastern Pennyroyal, the Bluegrass, and the western section of the Mountains. This pattern shows a considerable similarity to the pattern of tobacco acreage (Map 94).

Tobacco fields require large quantities of fertilizer. The yield is high and a few acres usually bring in considerable cash, making the investment in fertilizer highly profitable. Kentucky's fertilizer consumption for the year ending June 30, 1975, totalled 880,957 tons. Dry bulk fertilizer accounted for 50 percent of the market, dry bagged fertilizer for 25 percent, and liquid made up the remainder.

TRACTORS

MAP 110

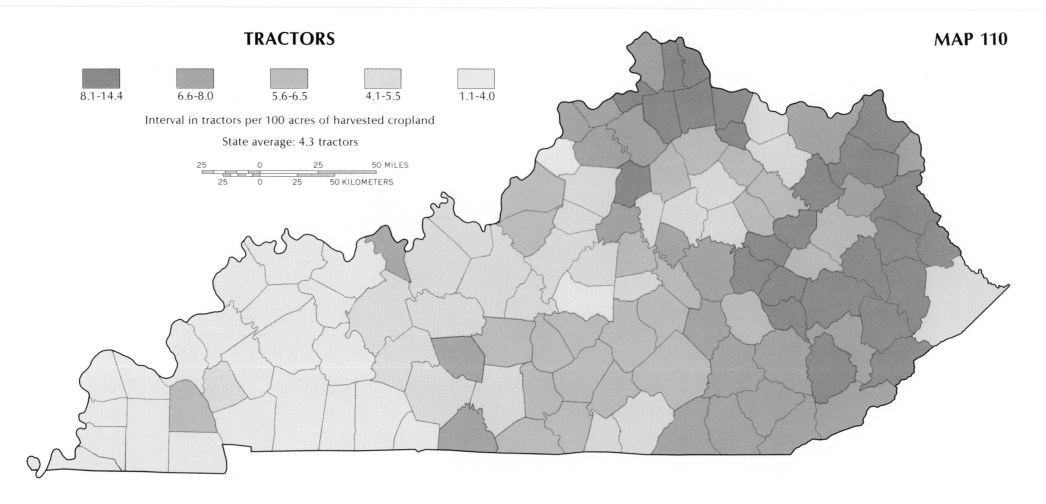

8.1-14.4 6.6-8.0 5.6-6.5 4.1-5.5 1.1-4.0

Interval in tractors per 100 acres of harvested cropland

State average: 4.3 tractors

25   0   25   50 MILES
25   0   25   50 KILOMETERS

Modern farms have substantial investments in machinery. In the last decade the use of time- and labor-saving machinery has greatly increased in Kentucky because of the continually increasing size of farms and the decreasing number of farm workers. This machine is harvesting bluegrass seed.

The pattern of tractors per 100 acres of harvested cropland (**Map 110**) shows a higher concentration in the Mountains and in the northern Kentucky counties. This reflects the many small tractors in use on relatively limited acres of cropland in these areas. In contrast, the large farms in central and western Kentucky, especially those producing soybeans, corn, wheat, and hay, use larger machines and more auxiliary sources of power. Tractors and other machinery represent a substantial capital investment on large farms. Increased efficiency through the use of tractors and machinery has led to increases in crop yields while lessening the demand for hand labor. The average size of farms in Kentucky has increased as the number of tractors has increased. This is particularly true in western Kentucky.

MAP 111

TENANT-OPERATED FARMS

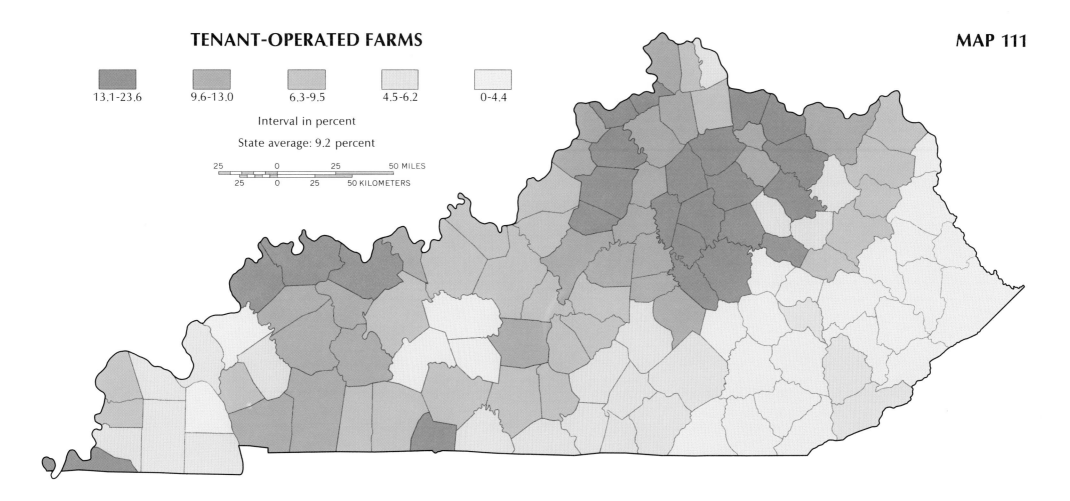

13.1-23.6 9.6-13.0 6.3-9.5 4.5-6.2 0-4.4

Interval in percent

State average: 9.2 percent

25   0   25   50 MILES

25   0   25   50 KILOMETERS

In 1940 about one-third of Kentucky's farms were operated by tenants. In the last three decades tenancy has decreased considerably. The state average in 1969 was 9.2 percent. Major concentrations of tenant-operated farms are found in the tobacco-growing areas of the Bluegrass and in the major grain-growing areas in western Kentucky (**Map 111**). Farm tenancy is low in the Mountains. Tenancy rates are highest in those areas where the value of farmland and farm buildings is highest.

Some tenants work for hourly wages while others receive a share of the crops. Usually a tenant farmer receives his house and garden rent-free. Often several members of a tenant family work part-time on the farm during the harvest season.

A small owner-operated farm in a valley in Breathitt County in the Mountains. Farm ownership is highest in eastern Kentucky while the percent of tenant-operated farms is highest in the Bluegrass Region and on some of the alluvial lands in the western part of the state, where farm values are highest.

Farm Income, Investment, & Ownership   147

VALUE OF FARM LAND AND BUILDINGS

MAP 112

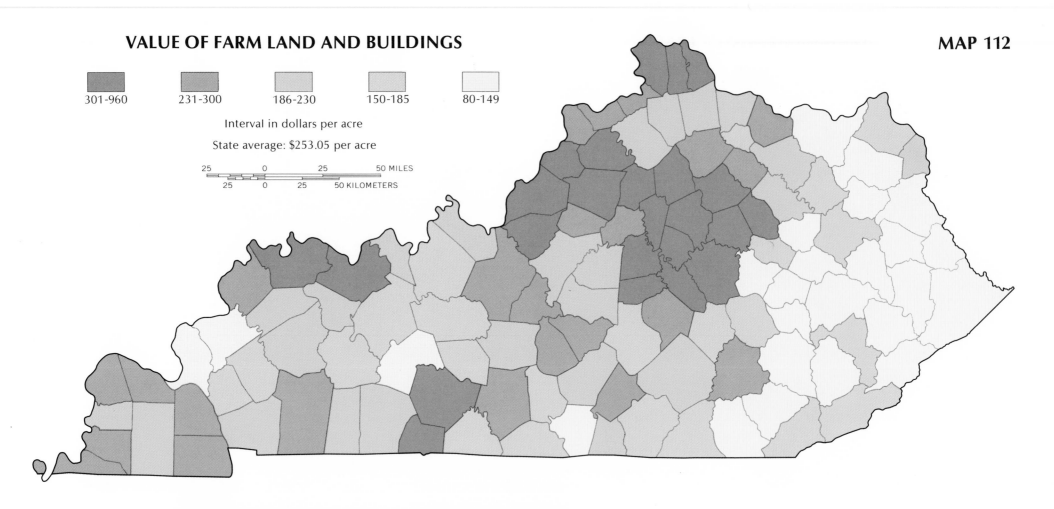

301-960   231-300   186-230   150-185   80-149

Interval in dollars per acre

State average: $253.05 per acre

25   0   25   50 MILES

25   0   25   50 KILOMETERS

A dairy farm for sale in the Eden Shale belt of Anderson County. Kentucky's farm values (average market value of land and buildings) have surged upward in the past few years. As of February 1, 1976, the average value per acre for farmland and buildings was $518 per acre, compared with $441 a year earlier and $394 in March 1974.

The highest value of farmland and buildings in Kentucky (**Map 112**) is in the agriculturally productive areas of the Bluegrass, in extreme northern Kentucky, and in Daviess, Henderson, Warren, and Simpson counties. The value of farmland has been appreciably increased in these areas by the urban encroachment of such cities as Louisville, Lexington, Covington, Bowling Green, Henderson, and Owensboro. The most extensive area of low-value farm property is in the Mountains, where poor soils and steep slopes prevail. The average value of farmland and buildings in 1975 was $441 per acre, an increase of 74 percent since 1969. The average value for the forty-eight contiguous states in that year was $354. Although the average value of farmland in Kentucky is less than that in Indiana, Ohio, Tennessee, Illinois, and North Carolina, the average for the Bluegrass region is one of the highest in the nation.

PART-TIME FARM OPERATION

MAP 113

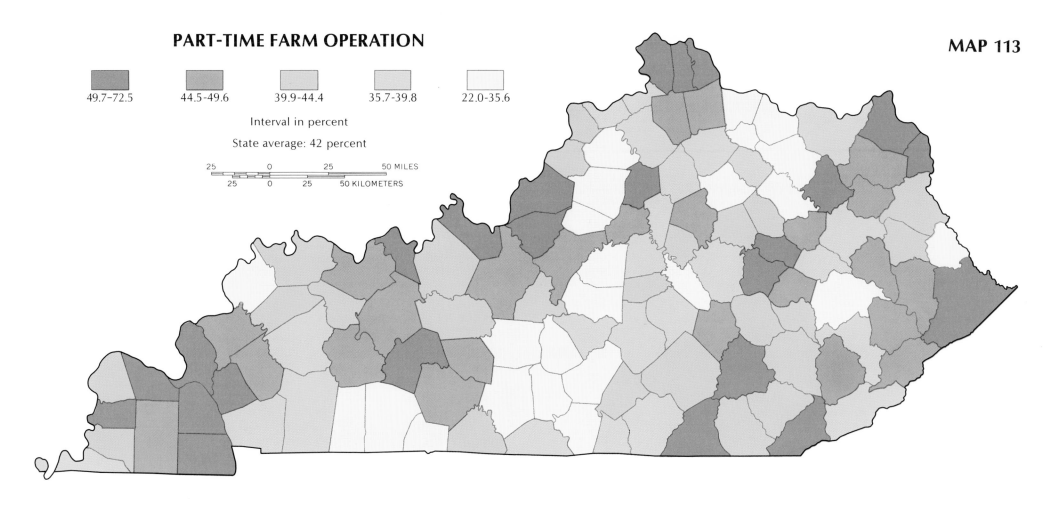

49.7-72.5 44.5-49.6 39.9-44.4 35.7-39.8 22.0-35.6

Interval in percent

State average: 42 percent

25   0   25   50 MILES
25   0   25   50 KILOMETERS

Part-time farm operations have increased markedly in recent years in Kentucky. Now nearly half the state's farm operations are in this category. The main concentrations of part-time farming are in areas adjacent to major urban centers and industrial or mining communities, where alternative sources of employment are available **(Map 113)**. From half to three-fourths of the farmers near Paducah, Louisville, Covington, and Ashland, and nearly half of those near Lexington, are part-time operators. Most of the part-time farmers in Kentucky hold full-time jobs in industry, government, or trade. Their farming operation usually consists of a small acreage in tobacco and truck garden produce, a few cattle and hogs, some acreage in corn for feed, and some pasture.

A part-time farmer outside Lexington in Fayette County. Nearly half the 124,000 farms in Kentucky in 1976 were part-time operations. The number of Kentucky farms has leveled off in the 1970s after a steady decline for several decades. The leveling off is due in part to the increasing number of part-time farmers near suburban areas.

Farm Income, Investment, & Ownership 149

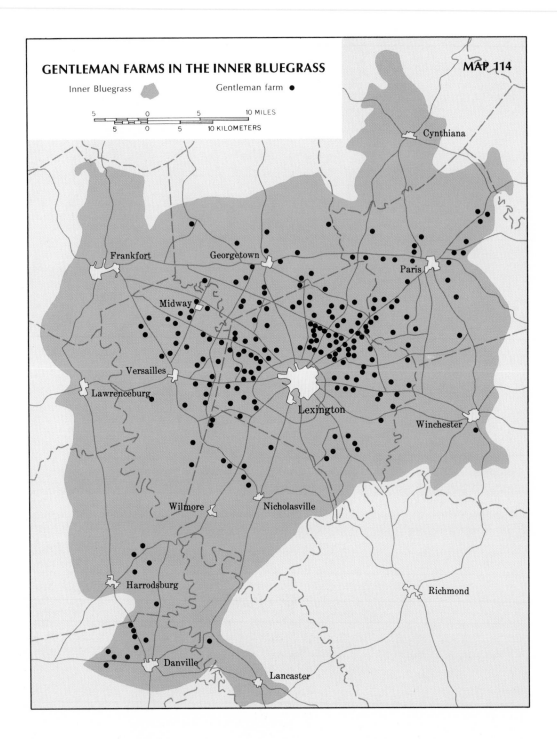

GENTLEMAN FARMS IN THE INNER BLUEGRASS

Inner Bluegrass   Gentleman farm ●

5   0   5   10 MILES
5   0   5   10 KILOMETERS

MAP 114

Cynthiana

Frankfort   Georgetown   Paris

Midway

Versailles

Lawrenceburg

Lexington

Winchester

Wilmore   Nicholasville

Harrodsburg

Richmond

Danville   Lancaster

Kentucky's Inner Bluegrass is the site of the largest concentration of gentleman farms in the United States **(Map 114)**. Gentleman farms emphasize the amenities of gracious rural living. For their owners farming is not an economic necessity nor does agriculture become their full-time vocation. Gentlemen farmers are independently wealthy and their rural interests are often an avocational experiment. Many but not all of Kentucky's gentlemen farmers engage in horse racing and breeding as their prime hobby, and consequently their establishments are often referred to as horse farms. Not all farms which keep or raise horses can be considered gentleman farms, however, and not all gentlemen farmers raise or race horses. Their interest may be in plant breeding experimentation, in purebred domestic beef cattle, or in more exotic animals such as bison or antelope. It is the combination of high social station and reputation, vast wealth, broad amplitude of interests and investments, and the habit of personifying these characteristics in material form on the landscape that defines the gentleman farmer.

The most constant feature of the gentleman farm is the creation and maintenance of an aesthetic farmscape. Set in perhaps the most beautiful of all rural landscapes, these farms are marked by long wooded driveways, ornamental entrances, limestone walls or stylish wooden fences, attractive ponds and creeks, and stately mansions. Most provide housing for the farm manager and many of the farm laborers on the property. Most buildings, including employee residences, are painted bright and distinctive colors, after the racing colors of the farm.

The largest cluster of gentleman farms is an extensive arc of greenbelt around the north side of Lexington extending outward from Fayette County into Bourbon, Scott, Woodford, and Jessamine counties. The second cluster is in Boyle and Mercer counties north of Danville. The strong historical legacy of large-scale plantation agriculture and aristocratic lifestyle possessed by early migrants and the beauty of the rolling karst topography have played an important role in influencing the location of gentleman farms in the Inner Bluegrass. These farms of the gentry have had a major impact on the image of Kentucky.

WILLIAM A. WITHINGTON and KARL B. RAITZ

Central Kentucky's gentleman farms are notable for their elegance and style. *Above:* The stately mansion at Darby Dan Farm, with the figure of a jockey serving as a hitching post, a traditional symbol of welcome. *Above right:* A creek flowing through the beautifully landscaped grounds at the C. V. Whitney farm. *Right:* A barn at Normandy Farm, designed after French provincial models. Although atypical, the gentleman farm is synonymous with life in the Bluegrass in the minds of many Americans.

XIII. RECREATION

Kentucky's recreational parks and state shrines (**Map 115**) have a wide variety of facilities for visitors. In fact Kentucky boasts one of the finest state park systems in the nation. The parks range in size from one-half acre to over 3,600 acres and are distributed throughout the state from the Appalachians to the Mississippi River. Sixteen parks with seasonal and year-round lodgings form an extensive recreational network. **Map 116** shows the origin of overnight guests at park lodges; visitation patterns at all parks and shrines are shown on **Map 117.**

Since 1954 tourist and travel activities have been among the most progressive sectors of the state's economy. Tourists from other states spent $575 million in Kentucky during 1974, and the retail market for out-of-state tourists is an important segment of the state's economy. Out-of-state tourist spending during 1974 increased 12 percent over 1973, and the increase from 1963 to 1974 was 161 percent. Spending by out-of-state tourists has been increasing at an average rate of 8.7 percent annually, as compared to 7.5 percent in the United States as a whole. Lodging, restaurant, and recreation services catering to the traveling public have increased 9.5 percent in Kentucky as compared to 8.6 percent in the nation.

The benefits from the tourist industry are not confined to private enterprise; they also make a significant contribution to the revenue

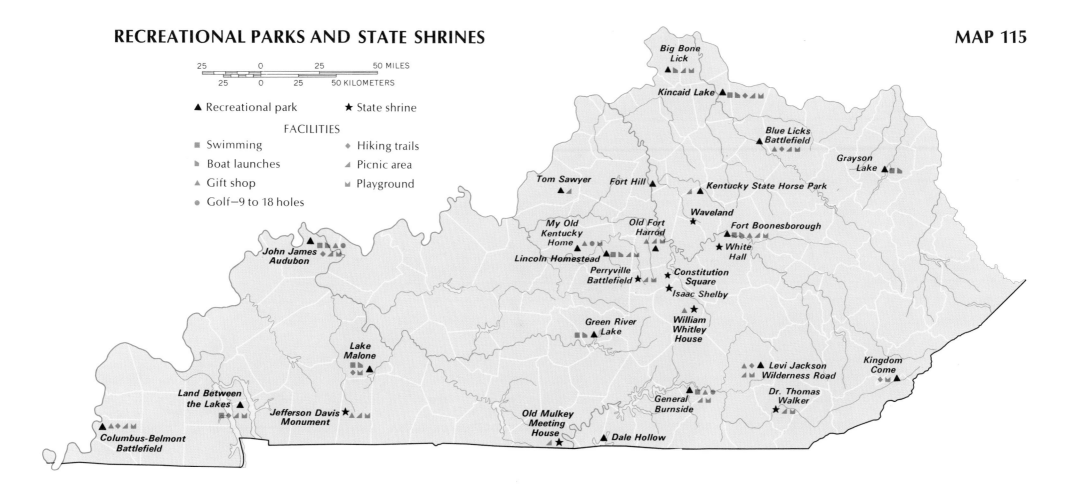

25 0 25 50 MILES
25 0 25 50 KILOMETERS

▲ Recreational park   ★ State shrine

FACILITIES

■ Swimming   ◆ Hiking trails
▶ Boat launches   ◢ Picnic area
▲ Gift shop   ◣ Playground
● Golf—9 to 18 holes

Big Bone Lick
Kincaid Lake
Blue Licks Battlefield
Grayson Lake
Tom Sawyer *Fort Hill* *Kentucky State Horse Park*
Waveland
My Old Kentucky Home *Old Fort Harrod* *Fort Boonesborough*
Lincoln Homestead *White Hall*
John James Audubon *Perryville Battlefield* *Constitution Square* *Isaac Shelby*
Green River Lake *William Whitley House*
Lake Malone *Levi Jackson Wilderness Road* *Kingdom Come*
Land Between the Lakes *Dr. Thomas Walker*
Jefferson Davis Monument *General Burnside*
Old Mulkey Meeting House *Dale Hollow*
Columbus-Belmont Battlefield

of state and local governments. The state government alone collected $262 million in gasoline, sales, and other business taxes from travelers during fiscal year 1974. In addition, $30 million were collected in local taxes. Tourists from other states pay nearly a quarter of these taxes. Kentucky's out-of-state tourists come from all over the eastern half of the United States, though more than 50 percent come from Illinois, Indiana, Ohio, and Michigan.

Probably the single best-known Kentucky focus of recreation for both out-of-state and in-state visitors is Mammoth Cave National Park in west-central Kentucky. Its many miles of underground limestone-soluble passageways are representative of a much broader area of karst topography in the state, where limestone layers predominate and where sinkholes, caves, and other limestone-solution features are common.

Eleven state parks are major attractors, with more than a million

Lake Cumberland in southeastern Kentucky (*opposite*), with a shoreline of over 1,200 miles, extends from Wolf Creek Dam on the Cumberland River through seven counties. As many as five million visitors a year come to this recreational area. *Left:* Colorful rock formations in Mammoth Cave National Park attract visitors from across the nation. The park also offers modern accommodations, campgrounds, hiking trails, and cruises on the Green River.

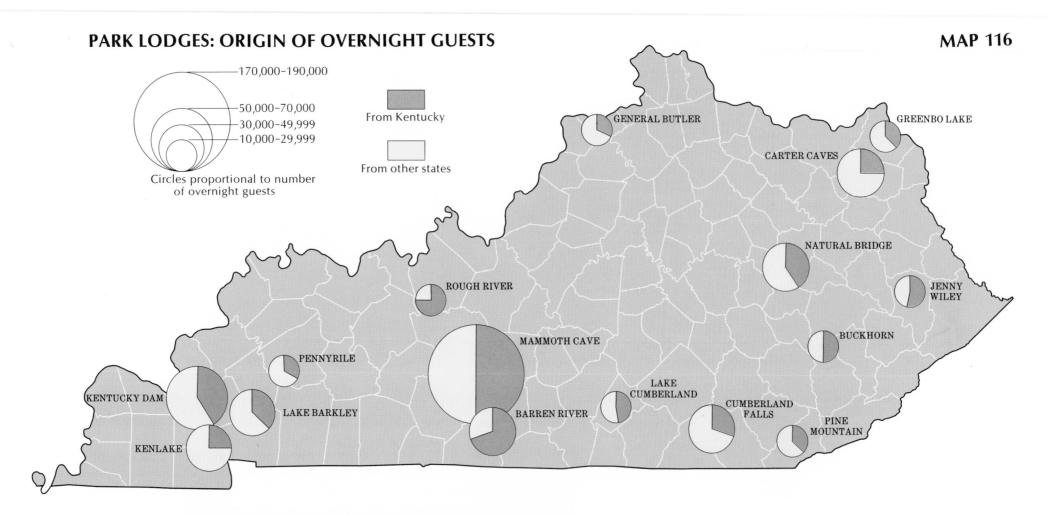

170,000–190,000

50,000–70,000

30,000–49,999

10,000–29,999

Circles proportional to number
of overnight guests

From Kentucky

From other states

GENERAL BUTLER

GREENBO LAKE

CARTER CAVES

NATURAL BRIDGE

JENNY WILEY

ROUGH RIVER

BUCKHORN

MAMMOTH CAVE

PENNYRILE

KENTUCKY DAM

LAKE CUMBERLAND

CUMBERLAND FALLS

LAKE BARKLEY

PINE MOUNTAIN

KENLAKE

BARREN RIVER

Lake Barkley Lodge, located on an embayment of the lake, provides modern accommodations for tourists visiting Lake Barkley State Resort Park. Named for Alben W. Barkley, U.S. vice-president (1949–1953), it is a major recreational area, offering boating, fishing, golf, and tennis. Out-of-state tourists to Kentucky's recreational areas spent over $650 million in 1975.

visitors each in 1973; two others had more than 880,000 visitors each (**Map 117**). Ten of these parks have lodges and associated recreational facilities. Five other parks with lodges attracted about a half million visitors each in 1973. Among the park lodges, Kentucky Dam Village in the Purchase, Mammoth Cave in the Pennyroyal, and Natural Bridge in eastern Kentucky are the most frequently visited in the state (**Map 116**). Lodges at Kenlake, Lake Barkley, Cumberland Falls, Barren River, and Carter Caves receive moderately large numbers of visitors. The remaining park lodges have less than 5 percent each of the total number of overnight guest registrations.

Two of the three most visited resort parks without lodges have natural settings as their principal attraction. These are Levi Jackson State Park in the southeastern Appalachian hill and mountain country in Laurel County, and Fort Boonesborough State Park on the

MAP 117

VISITORS TO PARKS AND SHRINES

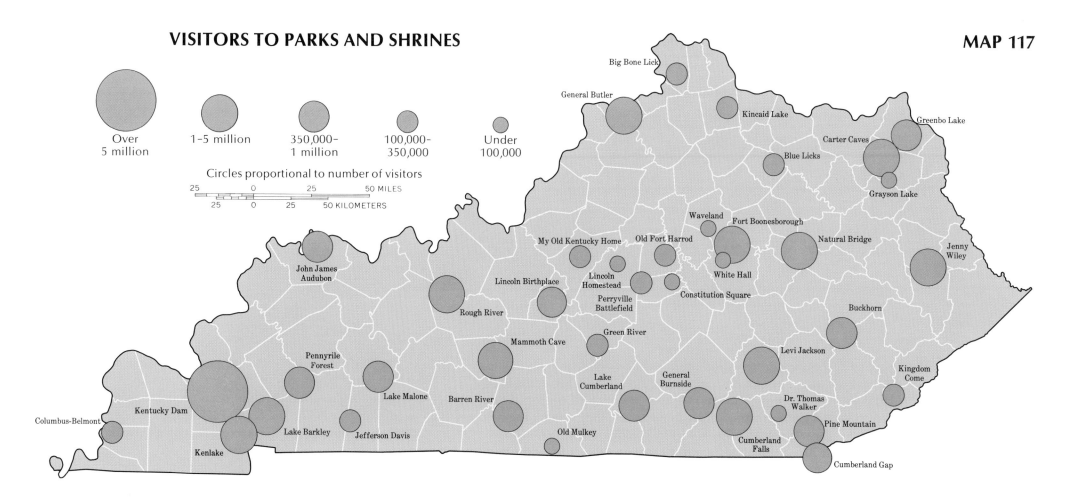

Over
5 million

1-5 million

350,000-
1 million

100,000-
350,000

Under
100,000

Circles proportional to number of visitors

25   0   25   50 MILES
25   0   25   50 KILOMETERS

Big Bone Lick

General Butler

Kincaid Lake

Greenbo Lake

Carter Caves

Blue Licks

Grayson Lake

Waveland

Fort Boonesborough

Natural Bridge

Jenny
Wiley

My Old Kentucky Home

Old Fort Harrod

John James
Audubon

Lincoln Birthplace

Lincoln
Homestead

White Hall

Perryville
Battlefield

Constitution Square

Buckhorn

Rough River

Green River

Mammoth Cave

Levi Jackson

Pennyrile
Forest

Lake
Cumberland

General
Burnside

Kingdom
Come

Lake Malone

Barren River

Dr. Thomas
Walker

Kentucky Dam

Columbus-Belmont

Pine Mountain

Lake Barkley

Jefferson Davis

Old Mulkey

Cumberland
Falls

Kenlake

Cumberland Gap

Kentucky River in northern Madison County, where Daniel Boone's fort has been recreated near its original site. The John James Audubon State Park near Henderson in the Western Coal Field focuses on a museum with Audubon's sketches and drawings of wildlife and only to a limited degree on surrounding recreational areas.

Kentucky's most frequented resort parks are associated with attractive natural, physical, or artificially created landscapes rather than with historic sites or the state's major urban centers. While three of the most visited state parks are not far from Paducah, their attraction lies primarily in the converging Tennessee and Cumberland rivers, the artificial lakes behind Kentucky and Barkley dams, and the Land Between the Lakes recreational area, developed by the Tennessee Valley Authority, rather than with Paducah. The seven parks centering on historic sites draw smaller numbers of visitors (**Maps 115 and 117**).

John James Audubon Museum near Henderson houses the paintings of the famed naturalist, who roamed this area of Kentucky from 1807 to 1819. Henderson is located on a major flyway for migrating birds. Out-of-state tourists vacationing in Kentucky parks contributed over $113 million in local, state, and federal taxes in 1975.

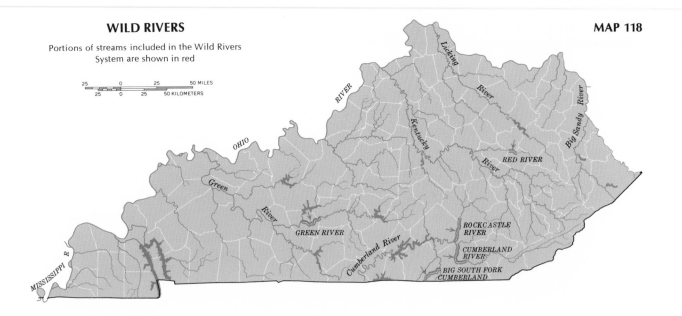

WILD RIVERS

Portions of streams included in the Wild Rivers
System are shown in red

MAP 118

In 1972 the Kentucky legislature passed a bill establishing a Wild Rivers System. The act designated certain streams for immediate inclusion in the system **(Map 118)** and presented the procedures and criteria for protecting and administering the system. The act provides for the preservation of scenic, ecological, and other values and for the proper management of recreational, wildlife, water, and other resources.

The portion of the Cumberland River from Summer Shoals to the backwater of Lake Cumberland, included in the Wild Rivers System, has traditionally been considered one of the clearest and finest streams in Kentucky. In recent years, however, its ecology has been severely altered as a result of surface disturbance in the headwaters, and large amounts of litter floating down from upstream sources have polluted the stream. Flowing through a region of enchanting scenery, the Cumberland River contains several uncommon species of fish which are threatened with extinction if the stream continues to deteriorate. It continues to support a fairly large black bass population, although not in the quantity that at one time made the Cumberland River one of the finest bass fishing streams in Kentucky.

The Rockcastle River from Kentucky Route 80 bridge to the backwater of Lake Cumberland is another stream designated as a wild river. It flows through an area of rugged wilderness and scenic beauty with limited accessibility. In addition to being a fine fishing stream, the Rockcastle serves as a major spawning area for the walleye and white bass fisheries of Lake Cumberland. Although the present pollution load is tolerable, it could become a threat if it is allowed to increase. The Big South Fork of the Cumberland River from the Tennessee line to Blue Heron is another stream included in the system.

A portion of the Green River downstream from Mammoth Cave National Park forms another segment of the Wild Rivers System. The stream flows through deep valleys and cavern areas in this section. With extensive subterranean courses and unique species of fish, this part of the Green River is one of the most remarkable streams in Kentucky. The area lying within Mammoth Cave National Park is of very high quality in terms of water purity, fish fauna, terrestrial wildlife, and scenic beauty. Many unique or rare species of fish and other wildlife, such as the Mammoth Cave blindfish, beaver, and muskellunge, inhabit the area.

The Red River from Kentucky Route 746 bridge to the mouth of Swift Camp Creek has also been designated a wild river. Until recently, the stream was clear, with only rubble and sand overlying a firm bedrock bottom, but a noticeable amount of finer sediment has begun to show up in recent years. Two possible sources of this silt are improper agricultural practices and lumbering. This deterioration of the water quality is not yet serious but it may become so in the future. In other respects, the quality of the water is excel-

FISHING WATERS

MAP 119

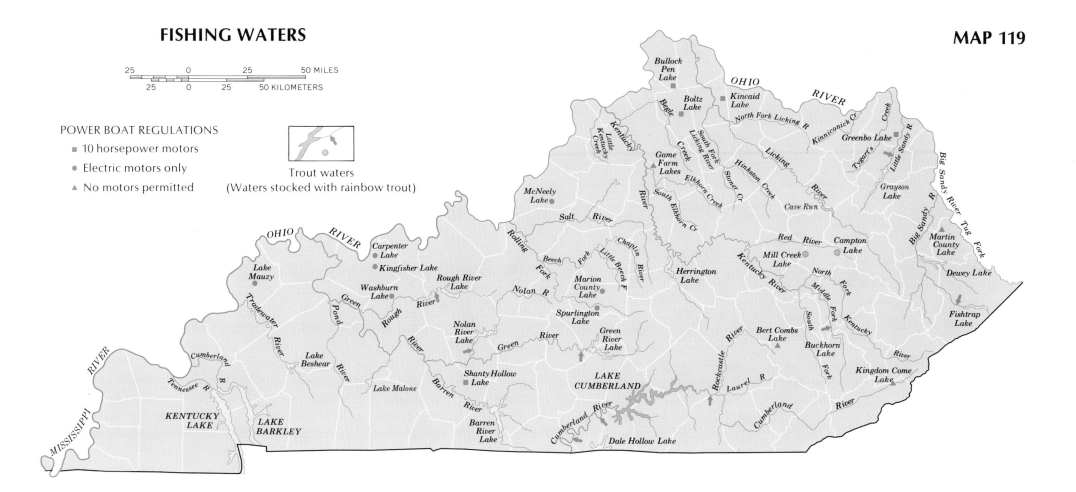

POWER BOAT REGULATIONS

■ 10 horsepower motors

● Electric motors only

▲ No motors permitted

Trout waters
(Waters stocked with rainbow trout)

lent: dissolved oxygen is high and the pH is near the point of neutrality. The fish fauna is marked to an exceptional degree by uncommon and endangered species. On the surrounding land, wildlife is abundant.

Kentucky's major rivers, notably the Ohio, Big Sandy, and Mississippi, together with a number of internal streams, provide primary recreational fishing **(Map 119)**. The mileage of waterways in Kentucky is one of the longest in the United States. Along some major streams and many minor ones as well, multipurpose projects have increased the diversity of uses, including recreational fishing. Kentucky Lake (Tennessee River), Lakes Barkley and Cumberland (Cumberland River), and a series of navigational improvement dams, some with hydroelectric capacities as well, exemplify the development and changes in use of rivers.

Fishing is literally a year-round activity at the sprawling Tennes-

see Valley Authority impoundments of Kentucky Lake and Lake Barkley. The annual crappie run in April and May yields slab-size fish ranging from one to three pounds. In late fall and through the early winter months, the tail waters provide excellent sauger fishing. White bass are plentiful in Barren River Lake. Good varieties and large yields of game and panfish are found in Rough River Lake. Lake Cumberland in southeastern Kentucky is one of the most scenic and productive fishing areas in the Commonwealth. Fishtrap Lake in Pike County offers bluegill and channel catfish. Buckhorn Lake contains a good population of white bass, providing excellent nighttime sport during the summer months. Rainbow trout are stocked on a monthly basis in several areas from spring through fall.

ALLAN J. WORMS, WILLIAM A. WITHINGTON, and WILFORD A. BLADEN

XIV. AIR AND WATER QUALITY

MAPS 120-123

Map 120 shows the air quality control regions of Kentucky, as delineated by the United States Environmental Protection Agency, and places in Kentucky exceeding the ambient air quality standards in 1974–1975. Paducah, Owensboro, Louisville's downtown area, and Newport's Mall area did not meet the primary ambient air quality standard for carbon monoxide. Part of Louisville, South Shore (south of Portsmouth, Ohio), Owensboro, Henderson, Newport, and Pikeville were among the places which had suspended particulates in the atmosphere exceeding the standard.

The Division of Air Pollution in the Kentucky Department for Natural Resources and Environmental Protection has authority, under regulations of the federal Clean Air Act of 1970, to ask business, industry, and government agencies for voluntary compliance with state directives during an air pollution alert. It can also impose extremely strict controls in the event of a designated emergency, when pollution levels are considered "immediately harmful" to people. In 1975 the Air Pollution Division called four pollution alerts. The first and longest began June 27 and lasted ten days. The other three were one or two days each. Since 1972, when the regulations first went into effect, Kentucky has never had anything more drastic than an alert. The culprit in each of the alerts in 1975 was ozone, a colorless, odorless gas formed by the action of sunlight and high temperatures on hydrocarbons and nitrogen dioxide. A stationary high pressure system bottles up high concentrations of ozone in stagnant air near the ground. Automobile exhaust fumes are considered a major source of ozone. During an alert motorists are asked to drive less and industries are asked to reduce emissions. Alerts are called county by county as air pollution monitors in major cities indicate the need.

The largest air pollution sources in Kentucky are the large power plants and industries located along the Ohio River and the automobile traffic associated with the major urban areas along the Ohio. Outside the Ohio Valley only a few places in Kentucky have significant amounts of suspended particulates in the air.

Air pollution sources in the Ashland area include steel, fertilizer, sulfuric acid, nickel, glass, and cement plants, gasoline refineries, a tannery, and a coke manufacturing operation, all of which release pollutants as smoke, dust, and gas. The location of the coke manufacturing plant and petroleum refineries is such that prevailing winds release the atmospheric pollutants over a large part of the community. Another source is the city dump, where open burning produces gaseous and particulate matter. In the Covington-Newport area several foundries, asphalt and scrap metal plants, dumps, breweries, and slaughter houses are the chief pollution sources. In the Louisville area numerous manufacturing establishments, including a slaughter house, a meat processing plant, and brewing and distilling industries, contribute contaminants into the atmosphere, especially suspended particulates and sulfur dioxide. In the Owensboro area large amounts of particulates emanate from concrete mix and concrete products plants. An oily mist released from a soybean processing plant provides the base for holding particulate matter which is deposited on top of it. A steel manufacturing plant, petroleum storage terminals, the municipal power plant, and woodworking plants release numerous air pollutants, including carbon and carbon monoxide, obnoxious amounts of wood dust, smoke, solvents, and odors.

In Henderson, concrete mix plants and sand dredging operations emit large quantities of particulate matter consisting of sand, ce-

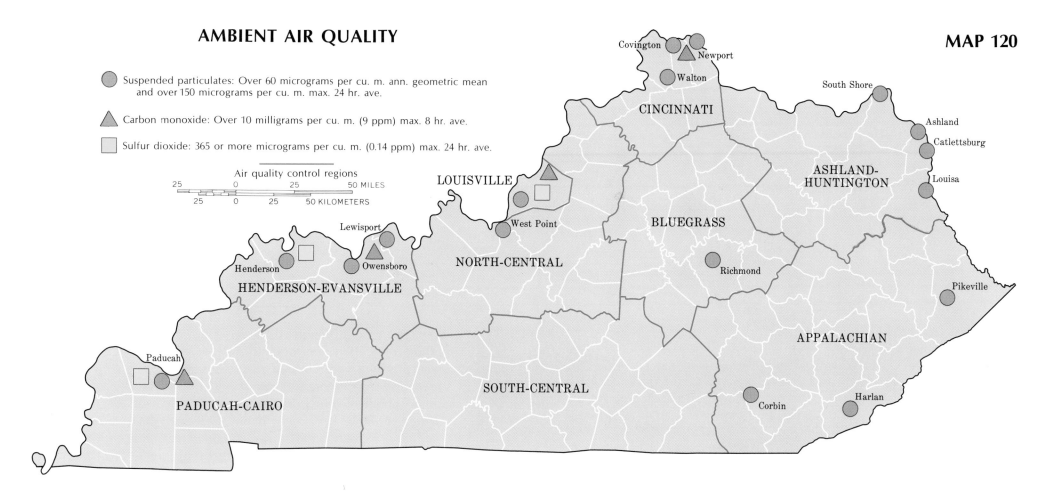

MAP 120

AMBIENT AIR QUALITY

Suspended particulates: Over 60 micrograms per cu. m. ann. geometric mean and over 150 micrograms per cu. m. max. 24 hr. ave.

Carbon monoxide: Over 10 milligrams per cu. m. (9 ppm) max. 8 hr. ave.

Sulfur dioxide: 365 or more micrograms per cu. m. (0.14 ppm) max. 24 hr. ave.

Air quality control regions

25   0   25   50 MILES
25   0   25   50 KILOMETERS

Covington   Newport
Walton
South Shore
CINCINNATI
Ashland
Catlettsburg
LOUISVILLE
ASHLAND-HUNTINGTON
Louisa
West Point
BLUEGRASS
Lewisport
NORTH-CENTRAL
Richmond
Henderson   Owensboro
HENDERSON-EVANSVILLE
Pikeville
APPALACHIAN
Paducah
SOUTH-CENTRAL
Harlan
PADUCAH-CAIRO
Corbin

ment, and gravel dusts. Hydrocarbons are released into the atmosphere from transfer and storage operations at petroleum terminals. In Paducah the burning of coal at a large hosiery mill releases particulate matter and sulfur dioxode into the atmosphere. An asphalt plant, a shoe factory, the city dump, a large railcar repair, demolition, and fabrication operation, meat packing houses, a veneer plant, machine shops, and petroleum terminals are other air pollution sources. Because the land in the area is fairly flat there is generally good air flow, except for occasional periods of stagnation and high humidity in summer.

Kentucky's private power plants and industries are on compliance schedules to meet the air quality standards. Holders of air pollution permits are required to submit reports on their emissions, to use approved control techniques, to follow a compliance schedule for reducing future emissions, and to follow prescribed emer-

gency procedures in pollution alerts. Federally owned plants, however, such as those operated by TVA at Paducah and Paradise, are among the largest air pollution sources in the state and the agencies operating these plants have not obtained air pollution permits. At issue are "constant" versus "intermittent" controls and a classic tug-of-war between the federal government and the state. The United States Environmental Protection Agency, which oversees the nation's drive toward clean air, backs the use of constant controls at power plants, and these are now the only kind permitted by the Kentucky Division of Air Pollution. Under the Clean Air Act the states are responsible for enforcing national clean air standards, with the Environmental Protection Agency approving each state's approach.

Although the stationary sources of pollution will be controlled, the problems of automobiles, which cause most of the smog prob-

MAP 121

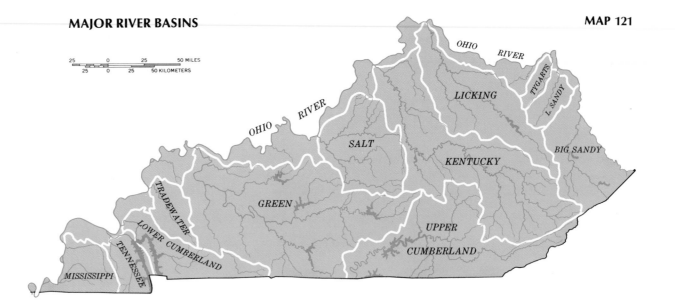

The Upper Cumberland River basin occupies a large area in southeastern Kentucky. This portion of it has been designated a "wild river." Kentucky's Wild Rivers System was established to protect the environmental quality of parts of the state's rivers which flow through wilderness areas of exceptional scenic beauty.

lem confronting Kentucky's larger cities, have not been resolved. The Air Pollution Control District in Louisville is dipping into land-use planning with several regulations. One deals with emission density zones, where clusters of polluting sources dictate that no more polluters can locate in a specific area. Another regulation levies restrictions on new public facilities which cause indirect sources of pollution, such as shopping and convention centers which attract many vehicles. Without proper traffic flow these areas could cause pockets of high air pollution. The emission density rule and the indirect sources regulations are the types of planning devices that will lead to lower levels of pollutants harmful to health.

The major drainage basins of Kentucky are, from east to west, the Big Sandy, Little Sandy, Tygarts, Ohio, Licking, Kentucky, Upper Cumberland, Salt, Green, Tradewater, Lower Cumberland, Tennessee, and Mississippi **(Map 121)**. About 97 percent of the total area of Kentucky drains into the Ohio River.

Kentucky's water resources as a whole are larger than all demands for water within the state in the foreseeable future. While this is true on the average, some places within the state suffer from water shortages during dry periods. A water shortage exists by definition when the combined demands for water in an area are not met by the minimum average historical stream flow calculated for seven consecutive days. By projecting future population and industrial and agricultural development, it is possible to predict future water needs.

◎ Existing   ● Projected

Butler
Falmouth
Cynthiana
Millersburg
Morehead
Frankfort
Owingsville
Mt. Washington   Taylorsville   Lexington
Shepherdsville   Lawrenceburg   Versailles   Winchester
Vine Grove   Lebanon Junction   Wilmore   Nicholasville   Clay City   West Liberty
Harrodsburg   Prestonsburg
New Haven   Lancaster   Jackson   Pikeville
Lebanon   Booneville   Elkhorn City
Liberty   Hazard
Mortons Gap   Greenville   Greensburg
Columbia
Harlan
Franklin

Within Kentucky many municipalities have current or projected water deficiencies (Map 122). The Kentucky River basin has thirteen municipalities with either a current or a projected water shortage; the Licking River basin has seven; the Salt and Green River basins have six each; the Big Sandy basin has three; and the Upper Cumberland has one. Continued growth in both population and economic development will intensify already existing problems and create new water shortage areas. There are two major ways to deal with the problem. First, low stream flows during the summer months can be compensated for by storing water during the winter months and releasing stored water during summer, a feat accomplished by using dams and reservoirs. Second, development in water-short areas can be discouraged in favor of development in areas with adequate water supplies.

While too little water at certain times of the year is a problem in many areas of the state, too much water at other times of the year has resulted in widespread property damage and deaths. Floods occur in Kentucky every year. Flash floods affect small areas in late spring or early summer, and floods affecting widespread areas occur in late winter. Flash floods are due to large amounts of precipitation in a short period of time, while the widespread floods of late winter are due to a combination of rainfall and increased runoff.

In terms of damage per square mile of floodplain, the Ohio River has by far the most severe flood problem in Kentucky. Large increases in flood damage are expected in the future along the Ohio, Big Sandy, and Kentucky rivers. Flood control projects provide part of the solution to the problem, but proper land-use planning and the control of floodplain development are equally important.

STREAMS AFFECTED OR POTENTIALLY AFFECTED BY MINE DRAINAGE

MAP 123

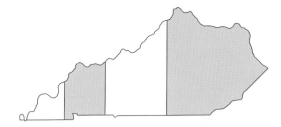

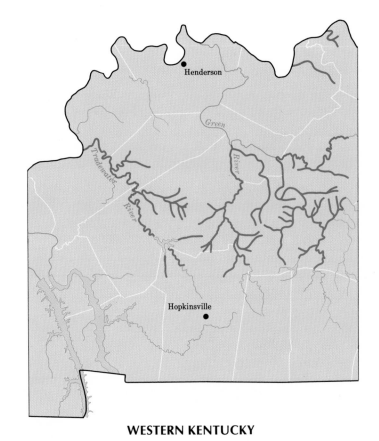

WESTERN KENTUCKY

EASTERN KENTUCKY

162

For the most part Kentucky's waters are of good quality. Major exceptions to this are streams affected by drainage from coal mines **(Map 123)**. If drainage cannot be prevented the acid present must be neutralized by alkaline river waters receiving the drainage. While for many streams the influence of mine drainage has been limited to an area close to the polluting source, some streams, such as the Tradewater River in western Kentucky and Carr Fork in eastern Kentucky, are acid for quite a distance below the source of pollution. The river basins most seriously affected are the Tradewater and Green River basins in western Kentucky and the Kentucky, Big Sandy, and Upper Cumberland River basins in eastern Kentucky.

Discharge of chemical water pollutants from surface mining generally occurs during periods of intense rainfall. During drier periods drainage water may often be trapped and form pools which con-

tain significant concentrations of chemical pollutants. These pools overflow during high rainfall periods and release the pollutants downslope and into a stream. Sometimes during periods of low precipitation the water in these pools percolates down through the overburden material and finds its way to a stream. These pollutants often adversely affect water quality by lowering pH values and increasing hardness by adding undesirable amounts of iron, manganese, calcium, sulfate, and other chemicals. Such pollution can be a primary cause of the destruction of plants, fish, and other forms of aquatic life. Other adverse effects are the deterioration of municipal and industrial water supplies, corrosion of bridges, of highway drainage installations, and of recreational and commercial boats, and lowering of water aesthetics in general.

WILFORD A. BLADEN and RICHARD I. TOWBER

XV. ADMINISTRATIVE ORGANIZATION

MAPS 124-127

EVOLUTION OF COUNTIES

Before Kentucky achieved statehood in 1792 it was part of Fincastle County, Virginia. Later known as Kentucky County, it was divided in 1780 into three counties: Fayette, Lincoln, and Jefferson. By June 1, 1792, when Kentucky became a separate state, the fifteenth to enter the Union and the first west of the Appalachians, its three counties had been subdivided into nine (**Map 124**). By 1818 Kentucky had fifty-nine counties (**Map 125**) and the Jackson Purchase, acquired in that year, added territory for eight more.

The Kentucky-Tennessee line was originally supposed to follow the 36° 30' parallel, but an error in surveying placed it north of this parallel. In 1820 the line was approved as surveyed, changing the southern boundaries of several Kentucky counties.

The growth of population created an administrative demand that led to the division of the state into 103 counties by 1855 (**Map 126**). Early Kentuckians felt that the size of a county should enable a man on horseback to get from his home to the county seat and back in one day, an attitude that accounts in part for the large number of counties. The subdivision of counties continued until 1912, when the creation of McCreary County brought Kentucky to its present 120 counties (**Map 127**).

The county is the basic unit of government in Kentucky. In most counties the judge, elected every four years, is the leading political figure. Other elected officials include treasurer, clerk, sheriff, jailer, and superintendent of schools. The county judge plays a dominant role in the fiscal and administrative affairs of the county in addition to trying juvenile and minor cases. The primary function of the sheriff is collection of taxes, but he also functions as a law enforcement officer. In many rural counties the superintendent of schools is the most influential elected official, after the county judge.

Barren County Courthouse in Glasgow, its architectural elegance symbolizing local pride. The county seat is an important political center, performing essential administrative functions.

MAP 124

COUNTIES, 1792

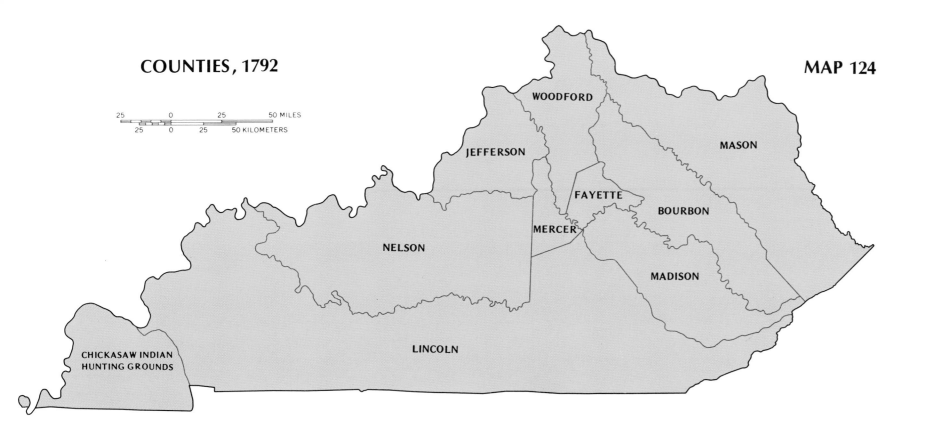

25 0 25 50 MILES
25 0 25 50 KILOMETERS

WOODFORD

JEFFERSON

MASON

FAYETTE

BOURBON

MERCER

NELSON

MADISON

LINCOLN

CHICKASAW INDIAN
HUNTING GROUNDS

COUNTIES, 1818

MAP 125

25 0 25 50 MILES
25 0 25 50 KILOMETERS

BOONE CAMPBELL

GALLATIN PENDLETON BRACKEN MASON LEWIS

HENRY FRANKLIN HARRISON NICHOLAS FLEMING GREENUP

JEFFERSON SHELBY SCOTT BOURBON BATH

WOODFORD FAYETTE CLARK MONTGOMERY

BULLITT JESSAMINE FLOYD

NELSON MERCER MADISON ESTILL

HENDERSON BRECKINRIDGE WASHINGTON GARRARD

UNION DAVIESS HARDIN LINCOLN CLAY

OHIO GRAYSON GREEN CASEY ROCKCASTLE

HOPKINS MUHLENBERG BUTLER ADAIR PULASKI

LIVINGSTON WARREN BARREN KNOX

CALDWELL

CHRISTIAN LOGAN WAYNE WHITLEY

THE JACKSON PURCHASE ALLEN CUMBERLAND

COUNTIES, 1855

25 0 25   50 MILES
25 0 25   50 KILOMETERS

MAP 126

BOONE
KENTON
CAMPBELL
GALLATIN
PENDLETON
BRACKEN
CARROLL
GRANT
TRIMBLE
OWEN
MASON
LEWIS
GREENUP
HENRY
HARRISON
NICHOLAS
FLEMING
CARTER
OLDHAM
SCOTT
BOURBON
BATH
LAWRENCE
JEFFERSON
SHELBY
FRANKLIN
MONTGOMERY
SPENCER
WOODFORD
FAYETTE
MORGAN
JOHNSON
ANDERSON
CLARK
POWELL
BULLITT
JESSAMINE
MEADE
NELSON
WASHINGTON
MERCER
MADISON
ESTILL
FLOYD
HENDERSON
HANCOCK
BRECKINRIDGE
HARDIN
BOYLE
GARRARD
BREATHITT
PIKE
DAVIESS
MARION
LINCOLN
OWSLEY
UNION
McLEAN
OHIO
GRAYSON
LARUE
GREEN
CASEY
ROCKCASTLE
CLAY
LETCHER
CRITTENDEN
HOPKINS
HART
TAYLOR
PERRY
LIVINGSTON
BUTLER
EDMONSON
ADAIR
PULASKI
LAUREL
KNOX
HARLAN
CALDWELL
MUHLENBERG
WARREN
BARREN
RUSSELL
McCRACKEN
LYON
BALLARD
MARSHALL
CHRISTIAN
TODD
LOGAN
SIMPSON
ALLEN
MONROE
CUMBERLAND
CLINTON
WAYNE
WHITLEY
HICKMAN
GRAVES
TRIGG
FULTON
CALLOWAY

COUNTIES, 1912

25 0 25   50 MILES
25 0 25   50 KILOMETERS

MAP 127

BOONE
KENTON
CAMPBELL
GALLATIN
PENDLETON
BRACKEN
CARROLL
GRANT
TRIMBLE
OWEN
MASON
ROBERTSON
LEWIS
GREENUP
HENRY
HARRISON
NICHOLAS
FLEMING
CARTER
BOYD
OLDHAM
SCOTT
BOURBON
BATH
ROWAN
ELLIOTT
JEFFERSON
SHELBY
FRANKLIN
MONTGOMERY
LAWRENCE
SPENCER
WOODFORD
FAYETTE
MENIFEE
MORGAN
ANDERSON
CLARK
JOHNSON
BULLITT
JESSAMINE
POWELL
WOLFE
MARTIN
MEADE
NELSON
WASHINGTON
MERCER
MADISON
ESTILL
LEE
BREATHITT
FLOYD
HENDERSON
HANCOCK
BRECKINRIDGE
HARDIN
BOYLE
GARRARD
MAGOFFIN
PIKE
DAVIESS
LINCOLN
JACKSON
OWSLEY
KNOTT
UNION
WEBSTER
McLEAN
OHIO
GRAYSON
LARUE
MARION
TAYLOR
CASEY
ROCKCASTLE
LAUREL
CLAY
PERRY
LESLIE
LETCHER
CRITTENDEN
HOPKINS
BUTLER
HART
GREEN
PULASKI
CALDWELL
MUHLENBERG
EDMONSON
ADAIR
KNOX
HARLAN
LIVINGSTON
WARREN
BARREN
METCALFE
RUSSELL
McCRACKEN
LYON
CHRISTIAN
TODD
LOGAN
CUMBERLAND
WAYNE
WHITLEY
BELL
BALLARD
MARSHALL
SIMPSON
ALLEN
MONROE
CLINTON
McCREARY
CARLISLE
TRIGG
HICKMAN
GRAVES
FULTON
CALLOWAY

AREA DEVELOPMENT DISTRICTS

MAP 128

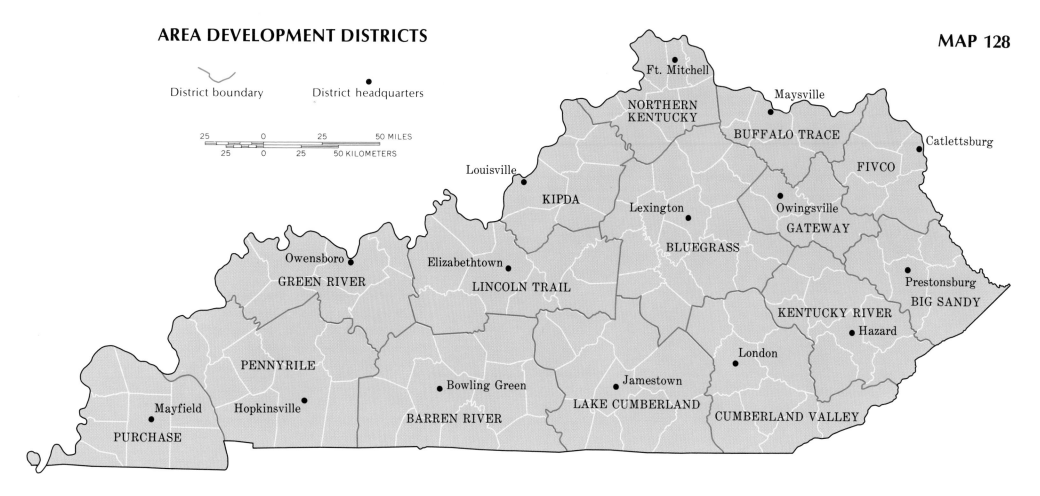

District boundary   District headquarters

25   0   25   50 MILES
25   0   25   50 KILOMETERS

ADMINISTRATIVE AND ELECTORAL DISTRICTS

MAPS 128-133

Area Development Districts (ADDs) are the keystones of the over-all development program in Kentucky **(Map 128).** These districts are multicounty units designed to bring together people who have common problems and goals because of similar geographical conditions and cultural backgrounds. The Kentucky Program Development Office (KPDO), established in 1968 and located in Frankfort, serves as coordinator for the fifteen districts and bridges the gap between state and local planners.

The area approach to planning and community development in Kentucky was first visualized in the late 1950s. A Kentucky Area Development Office was established in 1963 to provide liaison between local, state, and federal governments. The Appalachian Regional Development Act and the Public Works and Economic Development Act of 1965 gave added impetus to the establishment of planning districts. These acts required planning participation at the local level as a prerequisite for certification for federal funds. The purpose of the ADDs is to keep decision-making at the local government level where those making decisions are accessible to the average citizen. The combining of counties is intended to assure each area a sufficient scope and scale to organize and coordinate the many national and state programs. Each ADD is a nonprofit public corporation, registered with the Kentucky Secretary of State and governed by a board of directors which includes all county judges, one mayor for each county, and a number of citizens amounting to one fewer than the number of mayors and judges.

UNITED STATES CONGRESSIONAL DISTRICTS

MAP 129

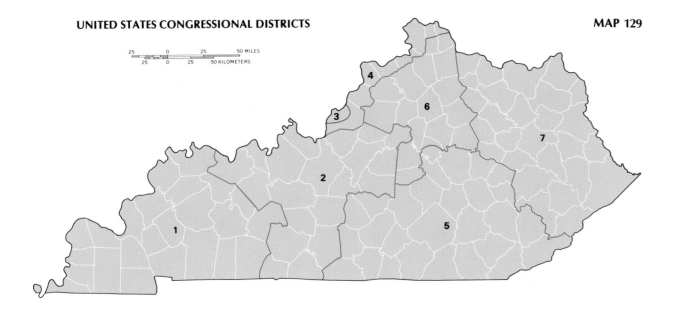

Kentucky has been subdivided in many ways according to administrative need. Among these subdivisions are the seven United States Congressional districts (**Map 129**), the forty-nine Circuit Court judicial districts (**Map 130**), the 100 General Assembly House of Representatives districts (**Map 131**), and the thirty-eight General Assembly Senate districts (**Map 132**).

The number of United States Congressional districts in a state depends upon the size of the state's population, and the boundaries of these districts are determined by the distribution of population, cutting across county lines where necessary to make districts of equal population. Each of Kentucky's Congressional districts elects a member to the House of Representatives. Boundaries of the seven districts were revised in 1972 on the basis of the census of 1970. The areas of equivalent population for election purposes have been so defined that the district boundaries are sometimes curious in terms of economic and social groupings.

Circuit Court judicial districts are formed by combining counties where necessary in an attempt to equalize case loads. Kentucky also has seven Court of Appeals districts.

Boundaries of General Assembly districts, like those of Congressional districts, cut across county lines in order to provide equal representation based on population, and are changed whenever a census shows inequality. The state legislature meets biennially in even-numbered years.

CIRCUIT COURT DISTRICTS

MAP 130

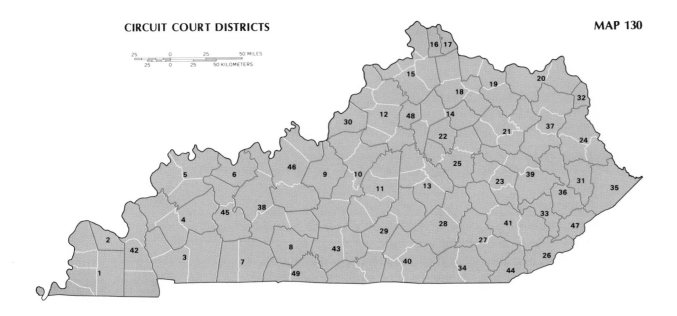

Circuit Court judges, elected for six-year terms, take the oath of office. They preside over courts dealing with both criminal and civil cases.

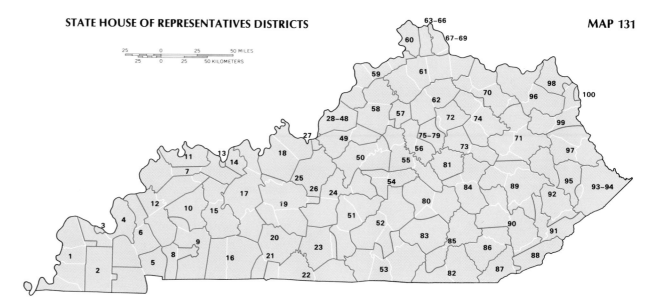

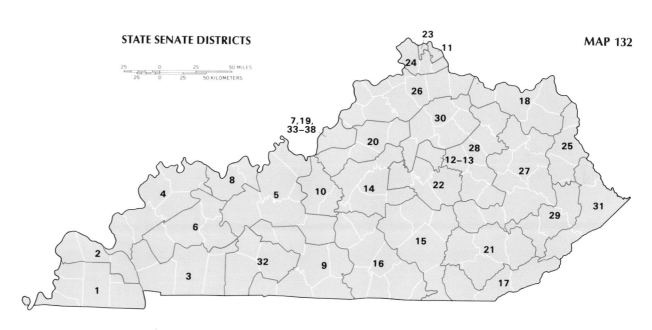

Above: The Kentucky State Capitol, overlooking the Kentucky River, was dedicated in 1910.

Left: Governor Julian M. Carroll addresses the 1976 joint session of the state legislature.

MAP 133

WET AND DRY COUNTIES

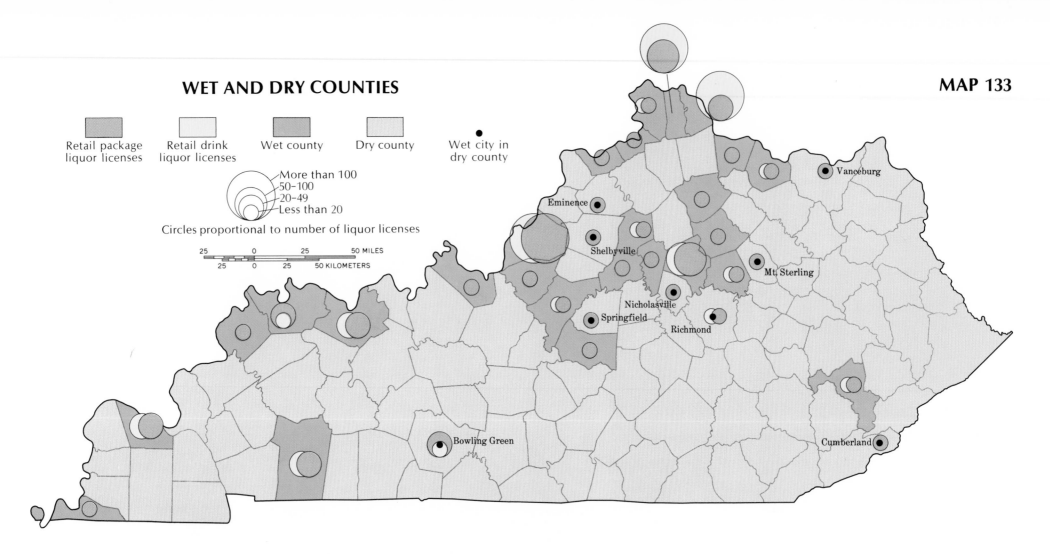

Retail package liquor licenses | Retail drink liquor licenses | Wet county | Dry county | • Wet city in dry county

More than 100
50–100
20–49
Less than 20

Circles proportional to number of liquor licenses

25 0 25 50 MILES
25 0 25 50 KILOMETERS

Vanceburg

Eminence
Shelbyville
Mt. Sterling
Nicholasville
Springfield
Richmond
Bowling Green
Cumberland

A liquor store at the border between Harrison County, which is "wet," and Scott County, which is "dry." In 1976, as a result of numerous local referendums, Kentucky had eighty-four dry counties, twenty-six wet counties, and ten counties that allowed the sale of alcoholic beverages only in certain cities.

Kentucky is a local option state. Each county decides by popular election if it will permit the sale of alcoholic beverages within its bounds. In 1974 twenty-six counties were "wet," eighty-five were "dry," and nine were divided (wet city in dry county) (Map 133). Several wet counties permit only package sales. Most of the larger urban centers have establishments with both retail package and retail drink liquor licenses. The numbers of retail drink and package establishments are about equal in most cities. Northern Kentucky is an exception in that package establishments predominate.

Most of the larger urban areas are wet, a major exception being Ashland. The absence of liquor sales in eastern, southern, and western Kentucky is attributable to the Anglo-Saxon Protestant heritage. Sunday sales of liquor are prohibited throughout the state. Kentucky has a "fair-trade" law that establishes minimum prices for liquors, but no liquor stores are state-owned.

MAP 134

NATIONAL GUARD UNITS AND MILITARY BASES

• National Guard unit ⬨ Military base

25 0 25 50 MILES
25 0 25 50 KILOMETERS

LOUISVILLE NAVAL
ORDNANCE PLANT

Carrollton
Ashland
Cynthiana
Carlisle
Olive Hill
Frankfort
Louisville
Lexington
HQ. LEXINGTON
BLUE GRASS
ARMY DEPOT
FT. KNOX
Bardstown Harrodsburg
Richmond
Henderson
Owensboro
Springfield
Ravenna
Prestonsburg
Elizabethtown
Danville
LEXINGTON
BLUE GRASS
ARMY DEPOT
Livermore
Jackson
Marion
Madisonville
Campbellsville
London
Central City
Paducah
Glasgow
Somerset
Bowling Green
Barbourville
Harlan
Hopkinsville
Russellville
Monticello
Middlesboro
Williamsburg
Hickman
Tompkinsville
FT. CAMPBELL

MILITARY AREAS

Kentucky has two large army bases: Ft. Knox, which occupies parts of Hardin, Meade, and Bullitt counties; and Ft. Campbell, which includes parts of Christian and Trigg counties and extends into Tennessee. Lexington-Blue Grass Army Depot has its headquarters near Lexington and also occupies about 14,000 acres in Madison County near Richmond. Louisville has a Naval Ordnance Depot. Thirty-eight Kentucky cities have National Guard units and some have army reserve units as well. The 100th Army Division, a training division, has its headquarters in Lexington. Military expenditures contribute significantly to Kentucky's economy and provide jobs for several thousand persons.

WILFORD A. BLADEN

MAP 134

A parade of Kentucky Air and Army National Guard units.

Over 1.15 million Kentucky voters went to the polls on November 2, 1976, to elect a president and vice-president, seven congressmen, and fourteen Court of Appeals judges, as well as local officials.

Above right: Rosalynn Carter during a visit to Lexington with a local official of her husband's campaign. A massive rural vote for Jimmy Carter in opposite ends of the state gave the 1976 Democratic presidential candidate a near landslide victory in Kentucky.

Right: Robert Dole, the 1976 Republican vice-presidential candidate, during a visit to Lexington. The Louisville and Cincinnati suburbs and southeastern Kentucky have long been strongly Republican, with central Kentucky a swing area but leaning Republican. In 1976 the Ford-Dole ticket carried only the 4th and 5th districts.

Opposite: Eligible Kentucky voters formed long lines to register in the fall of 1976 after an extensive voter registration drive.

XVI. VOTING PATTERNS

Of Kentucky's 3,218,706 inhabitants in 1970, 2,105,269 (65.4 percent) were of voting age (eighteen years and over). More than one-third (35.8 percent) of the voting-age population was in five counties: Jefferson (21.3 percent of the total), Fayette, Kenton, Campbell, and Daviess.

The last three gubernatorial elections have resulted in the evolution of a rather consistent pattern of voting behavior (Maps 135–138). Nearly two-thirds of Kentucky's 120 counties have voted consistently in terms of party preference; the others maintained a variable record, the bulk of them being in the northeast. In general, three areas are prominent as Democratic strongholds: the coal-rich counties of eastern Kentucky, the Outer Bluegrass, and western Kentucky. The Republican core consists of fifteen counties in southeastern Kentucky, most of which have historically returned Republican majorities of 60 percent or more in gubernatorial elections. A secondary Republican stronghold, on the eastern margins of the Western Coal Field, consists of Butler, Edmonson, and Grayson counties. More important in absolute votes, however, are Jefferson and Fayette counties, which have rather consistently voted Republican in the gubernatorial elections of recent decades, the election of 1975 (Map 137) being a notable exception.

In presidential elections (Maps 139–142) the pattern is different from that of gubernatorial elections: 75 percent of the counties offer a variable pattern of party preference. The Democratic core vanishes in the west and is reduced to only seven counties of the central Mountain area plus Carroll County. The Republican core is in the southeast. Crittenden County forms the only Republican stronghold in western Kentucky.

TERRY L. McINTOSH

MAPS 135–142

1967 GUBERNATORIAL ELECTION   MAP 135

Democrat   Republican

25   0   25   50 MILES
25   0   25   50 KILOMETERS

1975 GUBERNATORIAL ELECTION   MAP 137

Democrat   Republican

25   0   25   50 MILES
25   0   25   50 KILOMETERS

1971 GUBERNATORIAL ELECTION   MAP 136

Democrat   Republican

25   0   25   50 MILES
25   0   25   50 KILOMETERS

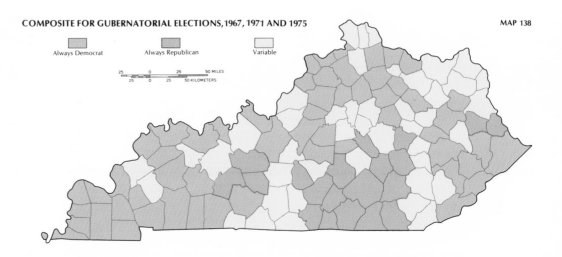

COMPOSITE FOR GUBERNATORIAL ELECTIONS, 1967, 1971 AND 1975   MAP 138

Always Democrat   Always Republican   Variable

25   0   25   50 MILES
25   0   25   50 KILOMETERS

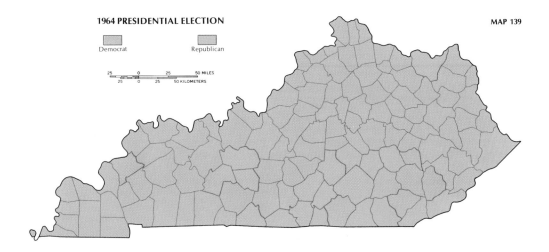

1964 PRESIDENTIAL ELECTION

MAP 139

Democrat   Republican

25 0 25 50 MILES
25 0 25 50 KILOMETERS

1972 PRESIDENTIAL ELECTION

MAP 141

Democrat   Republican

25 0 25 50 MILES
25 0 25 50 KILOMETERS

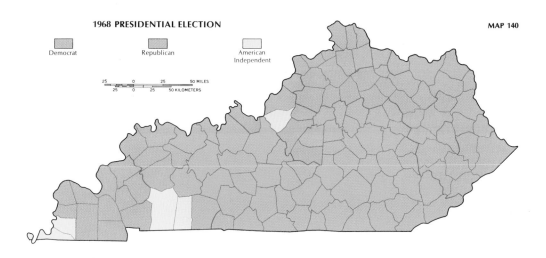

1968 PRESIDENTIAL ELECTION

MAP 140

Democrat   Republican   American Independent

25 0 25 50 MILES
25 0 25 50 KILOMETERS

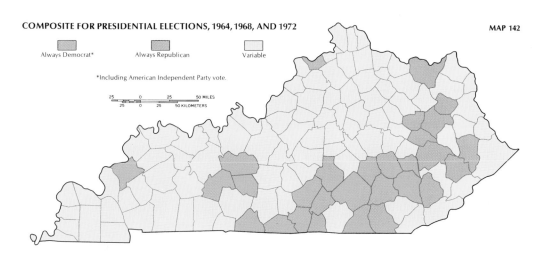

COMPOSITE FOR PRESIDENTIAL ELECTIONS, 1964, 1968, AND 1972

MAP 142

Always Democrat*   Always Republican   Variable

*Including American Independent Party vote.

25 0 25 50 MILES
25 0 25 50 KILOMETERS

Voting Patterns 175

Sources of Map Data and Selected References

MAP SOURCES AND REFERENCES

This section cites statistical and other sources used in preparing the maps and lists some additional useful references on the various topics. It is by no means an exhaustive bibliography on Kentucky. With few exceptions it leans toward current rather than retrospective bibliographic references.

Those interested in a bibliography of source materials on Kentucky may wish to consult *Inside Kentucky* (Frankfort: Department of Library Services, Kentucky Department of Education, 1974). They should also keep closely in touch with the publications of various state agencies. The *Economic Atlas of Kentucky* (Frankfort: Kentucky Agricultural and Industrial Development Board, 1952) adds some general background for comparison with the picture presented here, though the maps in that volume suffer from lack of specific detail and clarity. Readers interested in obtaining maps, aerial photographs, and geological publications for sale by the Department of Commerce Map Sales Office in Frankfort, Kentucky, should consult *Maps and Publications, 1976* (Frankfort: Kentucky Department of Commerce, 1976).

The mass of official literature on Kentucky issued by federal, state, and local government agencies is intimidating. Five major sources should be mentioned:

(1) U.S. Bureau of the Census, *Reports*, state volumes on Kentucky. For locating needed data and maps, see *Bureau of the Census Catalog: 1975 Annual* (Washington, D.C.: G.P.O., 1976).

(2) Reports of the Kentucky Geological Survey. See Preston McGrain and James C. Currens, *Bibliography of Industrial and Metallic Minerals in Kentucky through August 1973* (Lexington: Kentucky Geological Survey, 1975).

(3) Publications of the University of Kentucky Agricultural Experiment Station dealing with agricultural economics, forestry, livestock and dairying, population, soils and crops, and weather. See *Publications List* (Lexington: University of Kentucky College of Agriculture, Cooperative Extension Service).

(4) The numerous reports, drafts, and surveys issued by state agencies, such as the Kentucky Department of Commerce, the Legislative Research Commission, the Departments of Transportation, Agriculture, Parks, Health, Natural Resources, and Environmental Protection. These contain much information on economic, social, and physical aspects of the state. See annual volumes of *Checklist of Kentucky State Publications* (Frankfort: Division of Archives and Records).

(5) Various regional Area Development Districts and local planning agencies publish a variety of studies, maps, and resource inventories for their areas.

I. THE LAND OF KENTUCKY

Map 1, PHYSICAL FRAMEWORK: Based on the topographic sheets of Kentucky (scale 1:24,000). Cartographic work by Gyula Pauer.

For readers interested in a large map of Kentucky the most useful is: U.S. Geological Survey, "State of Kentucky" (scale 1:500,000), 27 by 57 inches, published in 1976. It shows highways, contours, shaded relief, principal cities and towns (with population indicated by size of letters), and national parks. It may be purchased from the U.S. Geological Survey, Reston, Virginia 22092.

Larger-scale topographic maps are also available on various scales:

(1) 1:24,000 (7.5-minute quadrangle series, 22 by 27 inches, covering approximately 59 square miles); published by the U.S. Geological Survey, 1948–1973. A total of 779 separate map sheets on this scale cover the entire state of Kentucky. See Charlotte Baumgarten, "Kentucky's King-size Map," *The Commonwealth* 2, no. 4 (July 1958): 4–7.

(2) 1:62,500 (15-minute quadrangle maps, 17 by 21 inches); published by the U.S. Geological Survey, 1947–1960. These modern, full-color map sheets of Kentucky, available for limited areas of the state, are beautiful, giving an enormous amount of information on land use, vegetation, and the cultural landscape.

The Kentucky Department of Commerce publishes state outline maps on various scales. These show county outlines along with county names, county seats, and streams. The Kentucky Department of Highways publishes annually the Official Highway Map of Kentucky on various scales. The most useful collection is *County Highway Maps: Bound Book 1974*, containing maps of all Kentucky counties on a scale of 1:125,000.

The Army Map Service has a topographic map series on a scale of 1:250,000 (12 sheets, each 24 by 34 inches), comprising 1° latitude and 2° longitude. The series, published between 1957 and 1969, provides coverage for the entire state. Prints of aerial photographs made for topographic mapping are available on scales varying from 1:14,000 to 1:44,000. Approximately twenty photos are required to cover a 7.5-minute topographic map sheet. Most Kentucky maps and aerial photographs can be purchased from the Map Sales Office, Kentucky Department of Commerce, Frankfort, Kentucky 40601.

Map 2, LAND REGIONS: Compiled by Thomas P. Field.

The first physiographic reconnaissance of Kentucky, made in 1838 by Dr. W. W. Mather, served to indicate six regions: (1) Cumberland Plateau or Mountains, (2) Bluegrass, (3) Knobs bounding the Bluegrass on the east, south, and west, (4) Pennyroyal or Mississippian Plateau, (5) Western Coal Field, and (6) Jackson Purchase. Mather did not use these regional names, which have now become standardized, nor did he define the boundaries. See W. W. Mather, "Report on the Geological Reconnaissance of Kentucky Made in 1838," Kentucky House of Representatives, *Journal* (1838–1839), pp. 239–78. Mather shows the Knobs and the Muldraughs Hill escarpment of the Knobs as the boundary of the Bluegrass. Likewise the outline of the Dripping Springs escarpment which impinges on the southern and southwestern borders of the Western Coal Field is portrayed as the outer limit of that region. See Richard F. Flint, "Natural Boundaries in the Interior Low Plateau Physiographic Province," *Journal of Geology* 36 (1928): 451–57.

A geographical account of these regions is given in P. P. Karan, ed., *Kentucky: A Regional Geography* (Dubuque, Iowa: Kendall/Hunt Publishing Co., 1973). For a detailed treatment of these Kentucky regions, readers should consult the six regional monographs written in the 1920s by prominent American geographers and published by the Kentucky Geological Survey under the direction of Dr. Willard R. Jillson, State Geologist. These monographs are: Darrell Haug Davis, *Geography of the Jackson Purchase* (Frankfort, 1923); idem, *Geography of the Kentucky Mountains* (Frankfort, 1924); idem, *Geography of the Blue Grass Region* (Frankfort, 1927); Wilbur Greeley Burroughs, *Geography of the Western Coal Field* (Frankfort, 1925); idem, *Geography of the Kentucky Knobs* (Frankfort, 1926); Carl Ortwin Sauer, assisted by John B. Leighly, Kenneth C. McMurry, and

Clarence W. Newman, *Geography of the Pennyroyal* (Frankfort, 1927). This series of studies, although nearly fifty years old, constitutes the only exhaustive study done of a single American state by professional geographers. Despite their age these regional surveys serve as models for other states. See also Darrell H. Davis, "A Study of the Succession of Human Activities in the Kentucky Mountains," *Journal of Geography* 29 (1930): 85–100; and Ellen Churchill Semple, "The Anglo-Saxons of the Kentucky Mountains," *Bulletin of the American Geographical Society* 42 (1910): 561–94.

For a discussion of regional geomorphology see Herbert Lehmann, *Zur Morphologie der Mitchellplain und der Pennyroyalplain in Indiana und Kentucky*, Deutscher Geographentag Bad Godesberg, October 1967, Tagungsbericht und Wissenschaft-liche Abhandlungen, Verhandlungen des Deutschen Geographentages, 36 (1969): 359–68; Armin Kohl Lobeck, *The Geology and Physiography of the Mammoth Cave National Park*, Kentucky Geological Survey Report, Series 6, pamphlet 21 (Frankfort, 1928); Samuel N. Dicken, "Kentucky Karst Landscapes," *Journal of Geology* 43 (1935): 708–28; idem, "The Kentucky Barrens," *Bulletin of the Geographical Society of Philadelphia* 33 (1935): 42–51; Franz-Dieter Miotke and Arthur N. Palmer, *Genetic Relationship between Caves and Landforms in the Mammoth Cave National Park Area: A Preliminary Report*, Varoffentlichung aus dem Geographischen Institut der Technischen Universitat Hannover (Wurzburg: Bohler, 1972).

II. EXPLORATION AND EARLY SETTLEMENT

Map 3, THE INDIAN AND EXPLORER TRAILS: Compiled by James E. Queen. Based on Thomas P. Field, *Map of Kentucky and the Southwest Territory, 1794*, Kentucky Study Series, no. 7 (Lexington: Department of Geography, University of Kentucky, 1966).

Additional information on the topic can be found in W. E. Myer, *Indian Trails of the Southeast*, Bureau of American Ethnology, 42nd Annual Report (Washington, D.C.: Government Printing Office, 1925); G. G. Clift, ed., *Guide to Kentucky Historical Highway Markers* (Frankfort: Kentucky Historical Society, 1969); J. S. Johnston, *First Explorations in Kentucky* (Louisville: J. P. Morton Co., 1898); August C. Mahr, "Shawnee Names and Migrations in Kentucky and West Virginia," *Ohio Journal of Science* 60 (1960): 155–64; Willard R. Jillson, *Pioneer Kentucky: An Outline of Its Exploration and Settlement . . .* (Frankfort: State Journal Co., 1934).

III. POPULATION CHARACTERISTICS

Tables 1 and **2:** BASED ON U.S. BUREAU OF THE CENSUS, *U.S. Census of Population, 1970:** *Number of Inhabitants, U.S. Summary*, tables 8, 9, and 18.

Table 3: Based on *1970 Census: Federal-State Cooperative Programs for Population Estimates*, Current Population Report Series, no. 126 (June 1975), p. 26.

Map 4, POPULATION DISTRIBUTION: Adapted from a population map prepared by John Fraser Hart and published in P. P. Karan, ed., *Kentucky: A Regional Geography* (Dubuque, Iowa: Kendall/Hunt Publishing Co., 1973), p. 8. The statistics on the number of inhabitants of Kentucky, its counties, county subdivisions, incorporated and unincorporated places, and other areas are given in *1970 Census: Number of Inhabitants, Final Report, PC(1)-A19 Kentucky* (1971).

Map 5, POPULATION CHANGE, 1950 TO 1960: Compiled by Ronald D. Garst. Based on data in U.S. Bureau of the Census, *County and City Data Book: Ken-

*Hereafter cited as *1970 Census*. All volumes were published by the Government Printing Office, Washington, D.C. Unless otherwise indicated, publication date was 1972.

tucky: A Statistical Abstract Supplement* (Washington, D.C.: G.P.O., 1962).

For a discussion of population changes in Kentucky, see George A. Hillery, *Population Growth in Kentucky, 1820–1960*, University of Kentucky Agricultural Experiment Station, Bulletin 705 (Lexington, 1966); and Gordon F. De Jong, *The Population of Kentucky: Changes in the Number of Inhabitants, 1950–60*, University of Kentucky Agricultural Experiment Station, Bulletin 675 (Lexington, 1961).

Map 6, POPULATION CHANGE, 1960 TO 1970 and **Map 7,** POPULATION AGE STRUCTURE, AGES 18–64: Compiled by Ronald D. Garst. Based on demographic statistics in *1970 Census: General Population Characteristics, Final Report PC(1)-B19 Kentucky* (1971).

Map 8, NATIVE POPULATION and **Map 9,** MIGRANT POPULATION: Compiled by Ronald D. Garst. Derived from *1970 Census: Social and Economic Characteristics: Kentucky*.

Maps 10, 11, and **12** were compiled by Stephen E. White.

Map 10, PRINCIPAL MIGRATION STREAMS: Based on data in *1970 Census: Migration between State Economic Areas*.

Map 11, IN-MIGRATION ATTRACTION OF CITIES, 1965–1970: Based on *1970 Census: Detailed Characteristics, Final Report PC(1)-D19 Kentucky*, data on migration and population. Attraction values for each city are defined as the number of in-migrants divided by the 1970 population. Biases created by armed services and college student migrations are eliminated.

Map 12, RESIDENTIAL PREFERENCE PATTERN: Based on a mail questionnaire survey; published originally in Stephen E. White "Residential Preference and Urban In-Migration," *Proceedings of the Association of American Geographers* 6 (1974): 47–50. Residential preference values were computed as an average of all the values granted a city by all respondents with sufficient knowledge, as follows: very undesirable = 0; undesirable = 25; fair = 50; desirable = 75; and very desirable = 100. Hometown evaluations were included in the average. A statewide residential preference pattern was mapped by assigning preference values at each city's location and plotting isopleths that evenly divided the range between the lowest and the highest preference values.

For migration from eastern Kentucky, see Harry K. Schwarzweller, James S. Brown, and J. J. Mangalam, *Mountain Families in Transition: A Case Study of Appalachian Migration* (University Park: Pennsylvania State University Press, 1971); Wayne T. Gray, "Population Movements in the Kentucky Mountains," *Rural Sociology* 10 (1945): 380–86.

Map 13, BLACK POPULATION and **Map 14,** CHANGES IN BLACK POPULATION, 1960–1970: Compiled by Karl B. Raitz. Based on maps and statistics in A. Lee Coleman and Doug I. Kim, *The Negro Population of Kentucky: Status and Trends, 1970*, University of Kentucky Agricultural Experiment Station, Bulletin 714 (Lexington, 1974).

See also Howard W. Beers and Catherine P. Heflin, *The Negro Population of Kentucky*, University of Kentucky Agricultural Experiment Station, Bulletin 481 (Lexington, 1946); A. Lee Coleman, Albert C. Pryor, and John R. Christiansen, *The Negro Population of Kentucky at Mid-century*, University of Kentucky Agricultural Experiment Station, Bulletin 643 (Lexington, 1956).

Map 15, MEMBERSHIP IN CHURCHES and **Map 16,** MEMBERSHIP IN FUNDAMENTAL DENOMINATIONS: Compiled by Richard W. Booth. Based on data in Douglas W. Johnson, Paul R. Picard, and Bernard Quinn, *Churches and Church Membership in the United States, 1971: An Enumeration by Region, State, and County* (Washington, D.C.: Glenmary Research Center, 1974).

IV. SOCIAL AND ECONOMIC PATTERNS

Map 17, SOCIOECONOMIC LEVELS: Compiled by Ronald D. Garst. Based on the results of a factor analysis of census data on twenty-four variables relating to income, education, and occupation in Kentucky counties in 1970. See *1970 Census: General Social and Economic Characteristics, Final Report PC(1)-C19 Kentucky.*

For an example of the use of factor analysis in delineating socioeconomic patterns, see John H. Thompson, Sidney C. Sufrin, Peter R. Gould, and Marion A. Buck, "Toward a Geography of Economic Health: The Case of New York State," *Annals of the Association of American Geographers* 52 (1962): 1-20.

Maps 18–21 were compiled by Phillip D. Phillips.

Map 18, POPULATION BELOW POVERTY LEVEL and **Map 19,** FAMILIES WITH INCOMES OVER $15,000: Based on data in *1970 Census: General Social and Economic Characteristics: Final Report PC(1)-C19 Kentucky,* tables 124 and 44.

Additional information on patterns of income can be found in H. K. Charlesworth, Janet S. Cabaniss, and K. David Epling, *Kentucky Personal Income Study, 1971 and 1972* (Lexington: University of Kentucky Office of Business Development and Government Services, 1974).

Map 20, INDEX CRIME RATE and **Map 21,** LAW ENFORCEMENT: Based on 1973 data, compiled from Kentucky State Police reports.

A discussion of the geographical aspects of crime is found in K. D. Harries, *The Geography of Crime and Justice* (New York: McGraw-Hill Book Co., 1974); and Phillip D. Phillips, "A Prologue to the Geography of Crime," *Proceedings of the Association of American Geographers* 4 (1972): 86–91.

Map 22 and **Figure 1,** OCCUPATIONAL STRUCTURE: Compiled by I. Shair. Based on 1974 data in *Kentucky: Deskbook of Economic Statistics, 1975* (Frankfort: Kentucky Department of Commerce, 1975), p. 32. **Table 4:** Based on 1970 and 1974 volumes of this annual publication, which contains the most frequently used statistical information on Kentucky.

Maps 23–25, EMPLOYMENT: PRIMARY, SECONDARY, AND TERTIARY ECONOMIC ACTIVITIES: Compiled by Ronald D. Garst. Based on data in *1970 Census: General Social and Economic Characteristics, Final Report DC(1)-C19 Kentucky.*

Map 26, AGE OF HOUSING; **Map 27,** MEDIAN HOUSING VALUES; and **Map 28,** HOUSING WITH COMPLETE PLUMBING FACILITIES: Compiled by Phillip D. Phillips. Based on data in *1970 Census: General Housing Characteristics, Final Report HC(1)-A19 Kentucky.*

Map 29, HOUSING ATTACHED TO A PUBLIC SEWER: Compiled by G. Wall. Based on data in *1970 Census: Detailed Housing Characteristics, Final Report HC(1)-B19 Kentucky.*

An easily available source of data on many topics depicted on maps in this and other sections of this atlas is Ralph J. Ramsey and Paul D. Warner, *Kentucky County Data Book, 1960 and 1970* (Lexington: University of Kentucky Cooperative Extension Service, 1974). This publication contains county data extracted from the U.S. census reports, dealing with population change, migration, residence, population age groups, education, employment, characteristics of farm operators and laborers, personal income, families below the poverty threshold, selected health indicators, crime indices, housing units and occupancy, age of housing units, and water, sewerage, plumbing, and kitchen facilities in housing units.

V. EDUCATIONAL DEVELOPMENT

Map 30, MEDIAN SCHOOL YEARS: Compiled by Phillip D. Phillips. **Map 31,** HIGHER EDUCATION: Compiled by Ronald D. Garst. Based on data from *1970 Census: General Social and Economic Characteristics, Final Report PC(1)-C19 Kentucky.*

Map 32, COUNTY OF ORIGIN OF STATE UNIVERSITY STUDENTS: Compiled by Christopher Boerner. Map and **Table 5** based on statistics in *Origin of Enrollments, Accredited Colleges and Universities, Fall Semester, 1973* (Frankfort:

Council on Public Higher Education, 1974); and *Kentucky College and University Enrollments, 1968* (Frankfort: Council on Public Higher Education, 1968).

VI. PATTERNS OF HEALTH CARE

Maps 33–37 were compiled by Richard I. Towber.

Map 33, PERSONS PER PHYSICIAN: Based on 1975 data from the Department of Community Medicine, University of Kentucky College of Medicine.

Map 34, PERSONS PER DENTIST: Based on 1975 data from the Department of Community Dentistry, University of Kentucky College of Dentistry.

Map 35, PERSONS PER NURSE: Based on data in *Selected Vital Statistics, 1972* (Frankfort: Kentucky Department for Human Resources, Management Systems Division, 1973).

Map 36, PERSONS PER ACUTE BED: Based on data from *Directory of Licensed Health Facilities and Health Services, January 1975* (Frankfort: Kentucky Department for Human Resources, Division for Licensing and Regulation, 1975).

Map 37, AVERAGE INFANT DEATH RATE, 1968–1972: Based on data from the 1968 through 1972 volumes of *Kentucky Vital Statistics* (Frankfort: Kentucky Department for Human Resources). This publication also contains information on cases of reportable diseases occurring in Kentucky for the year.

VII. TRANSPORTATION AND COMMUNICATION

Maps 38–41 were compiled by P. P. Karan and David Oakes.

Map 38, MAJOR RAIL ROUTES: Based on information from *Feasibility of East-West Rail Service in Kentucky,* Research Report no. 114 (Frankfort: Kentucky Legislative Research Commission, 1975).

For further information see Elmer G. Sulzer, *Ghost Railroads of Kentucky* (Indianapolis: Vane A. Jones, 1967); idem, "Kentucky's Abandoned Railroads," *Kentucky Engineer* 16 (1954): 9–20.

Map 39, NAVIGABLE WATERWAYS: Based on information from the U.S. Corps of Engineers, which has published navigation charts and maps for the Big Sandy, Ohio, Kentucky, Green, Tennessee, Cumberland, and Mississippi rivers.

Map 40, AIRPORT SERVICE RANGES: Based on a 1975 questionnaire survey of air passenger traffic and on *Kentucky Airport Directory, 1976–77* (Frankfort: Kentucky Department of Transportation, 1976).

Map 41, INTERCITY BUS ROUTES: Based on information from the 1975 edition of *Russell's Official National Motor Coach Guide* 48, no. 9, part 1, June 1976 (Cedar Rapids, Iowa: Russell's Guides, Inc.).

Map 42, HIGHWAY TRAFFIC FLOW: Compiled by William A. Withington. Based on *Traffic Flow Map, Kentucky, 1973* (Frankfort: Kentucky Department of Transportation, Bureau of Highways, Division of Planning, 1975).

Further information on Kentucky's transportation system can be found in *Planning for Highway Development in Kentucky* (Washington, D.C.: Roy Jorgensen and Associates, 1962); *Technical Report No. 3: Household Travel Survey* (Frankfort: Kentucky Department of Transportation, 1975); *An Analysis of Selected River Ports Developments in the Commonwealth of Kentucky* (Columbus, Ohio: Battelle Columbus Laboratories, 1973); *Kentucky Coal Transportation* (Lexington: Spindletop Research, Inc., 1975); *Kentucky Airport System Plan Report and Executive Summary, 1971: Technical Supplement, 1972* (Lexington: Spindletop Research, Inc., 1972); *Louisville Air Region Study Summary and Technical Supplements* (Los Angeles, California: Mann Daniel and Johnson and Mendenhall, 1973); and *Kentucky Air Commuter System Study* (Cincinnati, Ohio: Landrum and Brown, 1974).

Maps 43–45, CIRCULATION OF DAILY NEWSPAPERS and **Map 46,** NEWSPAPERS PUBLISHED OUTSIDE KENTUCKY: Compiled by Phillip D. Phillips. Based on data in *Circulation, 1972* (Northfield, Illinois: American Newspaper Markets, Inc., 1973).

Map 47, HOUSING WITH A TELEVISION: Compiled by Wilford A. Bladen. Based on data from *1970 Census: Detailed Housing Characteristics, Final Report HC(1)-B19 Kentucky.*

VIII. THE URBAN SYSTEM

Maps 48–54, EVOLUTION OF THE URBAN HIERARCHY: Compiled by William A. Withington and Phillip D. Phillips. Maps and **Table 6** based on data computed from U.S. census reports of 1800, 1840, 1870, 1910, 1930, 1950, and 1970. The method used in computing the urban hierarchy is an adaptation of one used by John Borchert in "American Metropolitan Evolution," *Geographical Review* 57 (1967): 301–32.

For an early study of urbanization see Darrell H. Davis, "Urban Development in the Kentucky Mountains," *Annals of the Association of American Geographers* 15 (1925): 92–99; and Charles Warren Thorthwaite, "Louisville: A Study in Urban Geography," Ph.D. dissertation, University of California, Berkeley, 1929.

Maps 55–58, COMMUTER SHEDS: Compiled by Phillip D. Phillips. Based on data from *Commuting Patterns of Kentucky Counties* (Frankfort: Kentucky Department of Commerce, 1973).

Map 59, AGE OF HOUSING; **Map 60,** INCOME; and **Map 61,** BLACK POPULATION: Compiled by Phillip D. Phillips. Based on data from *1970 Census: Census Tracts, Final Report PHC(1)-118 Louisville, Ky.-Ind. SMSA;* ibid., *-112 Lexington, Ky.;* ibid., *-44 Cincinnati, Ohio-Ky.-Ind.;* ibid., *-90 Huntington-Ashland, W. Va.-Ky.-Ohio.*

A valuable atlas containing twelve maps of Louisville depicting the spatial distribution of selected socioeconomic characteristics by census tract has been published by the U.S. Bureau of the Census and the Manpower Administration, *Urban Atlas Series: Tract Data for Standard Metropolitan Areas, GE80-4520, Louisville, Kentucky* (Washington, D.C.: G.P.O., 1974).

IX. MANUFACTURING AND TRADE

Map 62, MANUFACTURING EMPLOYMENT; **Map 63,** LINKAGE PATTERNS OF MANUFACTURING ESTABLISHMENTS; **Map 64,** BOURBON DISTILLERIES; and **Map 65,** TOBACCO WAREHOUSES AND MANUFACTURING PLANTS: Compiled by Wilford A. Bladen. Data for **Maps 62–64** from *Kentucky Directory of Manufacturers, 1974,* 17th ed. (Frankfort, Kentucky, Department of Commerce, 1975). Information on tobacco warehouses for **Map 65** was collected by field survey in the state.

Additional information on Kentucky industry can be obtained from various publications of the Kentucky Department of Commerce. These include *Industrial Resources: Louisville, Kentucky* (Frankfort, 1973); *Industrial Resources, Northern Kentucky* (Frankfort, 1973); and *Green River Basin Industrial Sites: Potential for Development in Kentucky* (Frankfort, 1967). For industrial growth in the Bluegrass Region see Dietrich M. Zimmer, *Die Industrialisierung der Bluegrass Region von Kentucky,* Heidelberg im Selbstverlag des Geographischen Instituts der Universitat, 1970.

Map 66, RETAIL SALES and **Map 67,** WHOLESALE TRADE: Compiled by Phillip D. Phillips. **Map 66:** Based on U.S. Bureau of the Census, *Census of Wholesale Trade, 1972: Area Series: Kentucky, WC 72-A-18* (Washington, D.C.: G.P.O., 1974).

Map 67 and **Table 7:** Based on data from U.S. Bureau of the Census, *Census of Retail Trade, 1972, Retail Trade: Major Retail Centers, Kentucky, RC 72-C-18* (Washington, D.C.: G.P.O., 1975).

X. GEOLOGY, MINERALS, AND ENERGY

Map 68, GENERALIZED GEOLOGY and **Map 69,** TECTONICS: Compiled by James E. Queen. Based on maps published by the United States Geological Survey, including new geologic map sheets published under the joint state-federal areal geologic mapping program. See Preston McGrain, "Economic Significance of Kentucky's Areal Geologic Mapping Program," *Transactions of the Kentucky Academy of Science* 24 (1964): 91–94; Wallace W. Hagan, *Progress Report of Kentucky Areal Geologic Mapping Program,* Kentucky Geological Survey, Special Publication 4 (1961), pp. 11–20. A comprehensive account of the state's geology is Arthur C. McFarlan, *Geology of Kentucky* (Lexington: University of Kentucky, 1943).

Maps 70–74 were compiled by Wilford A. Bladen.

Map 70, MINERAL RESOURCES: Based on information from the Kentucky Geological Survey.

Map 71, COAL MINING EMPLOYMENT AND WAGES and **Map 72,** SURFACE MINES IN EASTERN KENTUCKY: Based on data from the Kentucky Department of Commerce and on Kentucky Department of Mines and Minerals, *Annual Report, 1974* (Lexington, 1975).

For coal resources of eastern Kentucky see J. W. Huddle et al., *Coal Reserves of Eastern Kentucky,* U.S. Geological Survey, Bulletin 1120 (Washington, D.C.: G.P.O., 1963). For western Kentucky see David L. Hodgson, *Coal Reserves in the Upper Tradewater River Area, Western Kentucky* (Chattanooga: Tennessee Valley Authority, 1963). Surface mining in eastern Kentucky is examined in Harry Caudill, *My Land Is Dying* (New York: Dutton, 1972). Public reaction to surface mining in eastern Kentucky is examined by W. A. Bladen, "Perception of Environmental Problems in Coal Mining Areas of India and the United States," *National Geographer* 10 (1975): 1–8. The reclamation problem in western and eastern Kentucky is analyzed by Lee Guernsey, "The Reclamation of Strip Mined Lands in Western Kentucky," *Journal of Geography* 59 (1960): 5–11; and Jerry E. Green, "The Problem of Reclamation of Derelict Land after Coal Stripmining in Appalachia," *Southeastern Geographer* 9 (1969): 36–47.

Map 73, ELECTRIC GENERATING STATIONS and **Map 74,** OIL AND GAS: Based on 1975 data from the Kentucky Department of Commerce, Frankfort. Further information on the energy problem is contained in Kentucky Development Cabinet, *Energy Network, 1974–1975* (Frankfort, 1975).

XI. LAND USE AND PHYSICAL CHARACTERISTICS

Map 75, LAND USE WITHIN PHYSIOGRAPHIC REGIONS: Compiled by Wilford A. Bladen. Based on 1975 data from the Kentucky Department of Commerce.

Map 76, PHYSIOGRAPHIC REGIONS: Adapted from Armin Kohl Lobeck's physiographic diagram of Kentucky in *The Midland Trail in Kentucky: A Physiographic and Geologic Guide Book to U.S. Highway No. 60,* Kentucky Geological Survey Report, series 6, pamphlet 23 (Frankfort, 1930).

Map 77, PHYSIOGRAPHIC AND MAJOR SOIL ASSOCIATION AREAS: Compiled by Harry H. Bailey. Based on information from the University of Kentucky Agricultural Experiment Station.

Additional information on soils of Kentucky can be found in U.S. Soil Conservation Service, *Soil Survey Laboratory Data and Description of Some Soils of Kentucky,* Soil Survey Investigations, report no. 14 (Washington, D.C.: G.P.O., 1967); P. E. Karraker, *Soils of the Different Regions in Kentucky,* Kentucky Agricultural Experiment Station, circular 67 (Lexington, 1950); W. S. Ligon and P. E. Karraker, *A Key to Kentucky Soils,* Kentucky Agricultural Experiment Station, circular 64 (Lexington, 1949); and S. N. Dicken and H. B. Brown, *Soil Erosion in the Karst Lands of Kentucky: Physiographic Conditions Affecting Erosion and Land*

Use in Areas Underlain by Soluble Limestone, U.S. Department of Agriculture, circular 490 (Washington, D.C.: G.P.O., 1938).

Map 78, LAND USE SUITABILITY: Compiled by Gyula Pauer. Based on data collected from Kentucky Departments of Agriculture and Forestry and information gleaned from air photographs and satellite imagery.

Maps 79–88, CLIMATIC CHARACTERISTICS: Compiled by Karl B. Raitz. Maps and **Table 8:** Based on data from the state climatologist, Agricultural Experiment Station, University of Kentucky.

Additional information is found in "Climate of Kentucky," U.S. Department of Agriculture, *Yearbook of Agriculture, 1941*, pp. 884–93; S. S. Visher, *The Climate of Kentucky*, Kentucky Geological Survey Report, series 6, pamphlet 31 (Frankfort, 1929): 87–167; and Harry K. Hutter, "A Climatic Study: Lexington, Kentucky," *Ohio Journal of Science* 49 (1949): 221–29. Another valuable publication is Stephen S. Visher, *The Climatic Maps of the United States* (Cambridge: Harvard University Press, 1954), which contains 1,031 maps and diagrams dealing clearly and simply with every aspect of American climate.

XII. FORESTRY AND AGRICULTURE

Maps 89–95 and **97–114** were compiled by Karl B. Raitz.

Map 89, LAND IN COMMERCIAL FOREST: Based on information in U.S. Forest Service, Resource Bulletin NE-9 (Washington, D.C.: G.P.O., 1968); and *Kentucky Deskbook of Economic Statistics, 1975* (Frankfort: Kentucky Department of Commerce, 1975). Data on wood-based industries come from the Forest Product Utilization Section of the Kentucky Division of Forestry.

Maps 90–93, FARMLAND AND FARM TYPES; **Maps 94–99,** MAJOR CROPS AND PASTURELAND; **Maps 100, 102, 103,** and **105,** COMMERCIAL LIVESTOCK; and **Maps 107–113,** FARM INCOME, INVESTMENT, AND OWNERSHIP: Based on U.S. Department of Commerce, *1969 U.S. Census of Agriculture, Part 30, Kentucky, Sec. 1, Summary Data, Vol. 1, Area Reports* (Washington, D.C.: G.P.O., 1973). Data from the agricultural census taken in 1974 will not be available until 1977 or 1978. To make these maps as contemporary as possible, the data have been checked wherever possible with *Kentucky Agricultural Statistics, 1974–75* (Louisville: Kentucky Coop and Livestock Reporting Service, 1975). This annual publication does not provide county data on all aspects of agriculture mapped here, but includes important facts about coops, livestock, the dairy industry, farm income, and other facets of agriculture.

Map 96, SOYBEANS and **Figure 2,** VALUE OF AGRICULTURAL PRODUCTION: Compiled by William A. Withington. Based on data in *Kentucky Agricultural Statistics, 1974–75.*

Map 101, FLUID MILK SOURCE REGIONS and **Map 104,** LIVESTOCK SALES OUTLETS: Based on 1974 information from the University of Kentucky Agricultural Experiment Station.

For additional information see *Kentucky Dairy Industry Facts*, University of Kentucky Agricultural Experiment Station, Progress Report 156 (Lexington, 1970).

Map 106, THOROUGHBRED HORSE FARMS: Based on 1975 data supplied by Thoroughbred Breeders of Kentucky, Lexington.

Map 114, GENTLEMAN FARMS IN THE INNER BLUEGRASS: Based on data gathered by field survey. Eight farmscape criteria were used: driveway, ornamental entrance, fence, parkland landscape, mansion style and age, purebred livestock, farm color scheme, and employee residence. A farm fulfilling any five of these criteria was counted as a gentleman farm. See Karl B. Raitz, "Gentleman Farms in Kentucky's Inner Bluegrass: A Problem in Mapping," *Southeastern Geographer* 15, no. 1 (1976): 33–46.

Additional information on geographic aspects of Kentucky agriculture can be found in John Fraser Hart, "Abandonment of Farmland in Kentucky," *Southeastern*

Geographer 4 (1964): 1–10; idem, "Abandonment of Farm Land on the Appalachian Fringe of Kentucky," in *Melanges de Geographie: Physique, Humaine, Economique, Appliquee, Offerts a M. Omer Tulippe*, vol. 1, Geographie physique et geographie humaine (Gembloux: J. Duculot, 1967), pp. 352–60; J. Sullivan Gibson, "Land Economy of Warren County," *Economic Geography* 10 (1934): 75–98, 200–216, 268–87; W. A. Browne, "Dark-fired Tobacco Region of the North Highland Rim," *Economic Geography* 14 (1938): 55–67; Leonard S. Wilson, "Land Use Patterns in the Inner Bluegrass," *Economic Geography* 17 (1941): 287–96; Raymond E. Murphy, "Land Values in the Bluegrass and Nashville Basins," *Economic Geography* 6 (1930): 191–203; R. W. Johnson, "Land Use in the Bluegrass Basin," *Economic Geography* 16 (1940): 315–35; and J. Russell Whitaker, "The Development of the Tobacco Industry in Kentucky: A Geographical Interpretation," *Bulletin of the Geographical Society of Philadelphia* 27 (1929): 15–42.

XIII. RECREATION

Maps 115–119, RECREATION: Based on data from the Kentucky Department of Parks, Frankfort. **Map 115,** RECREATIONAL PARKS AND STATE SHRINES, **Map 118,** WILD RIVERS, and **Map 119,** FISHING WATERS: Compiled by Wilford A. Bladen. **Map 116,** PARK LODGES: ORIGIN OF OVERNIGHT GUESTS and **Map 117,** VISITORS TO PARKS AND SHRINES: Compiled by William A. Withington. The data refer to 1973, the latest year for which complete statistics are available.

Additional information on recreational aspects of Kentucky can be found in *Comprehensive Outdoor Recreation Plan for the Commonwealth of Kentucky* (Frankfort: Kentucky Department of Parks, 1973); *Kentucky Tourist Preferences* (Lexington: University of Kentucky Bureau of Business Research, 1962); *Survey of Historic Sites in Kentucky* (Lexington: Spindletop Research, Inc., 1971); Max Hunn, "State Parks of Kentucky," *Travel* 124 (1965): 23–27. For a discussion of visitation patterns in Kentucky, see Wayne L. Hoffman and G. H. Romsa, "Some Factors Influencing Attendance at Commercial Campgrounds: A Case Study," *Land Economics* 48 (1972): 188–90.

XIV. AIR AND WATER QUALITY

Maps 120–123 were compiled by Wilford A. Bladen.

Map 120, AMBIENT AIR QUALITY: Based on statistics in *Air Quality Data Summary, October 1974–September 1975* (Frankfort: Air Quality Section Technical Services Program, Division of Air Pollution, Kentucky Department for Natural Resources and Environmental Protection, 1975). Information on air pollution standards is given in *Kentucky Air Pollution Control Regulations* (Frankfort: Division of Air Pollution, Bureau of Environmental Protection, 1975).

Map 121, MAJOR RIVER BASINS: Based on topographic maps and materials in R. A. Krieger, R. V. Cushman, and N. O. Thomas, *Water in Kentucky* (Lexington: Kentucky Geological Survey, 1969).

Map 122, MUNICIPALITIES WITH WATER SUPPLY PROBLEMS and **Map 123,** STREAMS AFFECTED OR POTENTIALLY AFFECTED BY MINE DRAINAGE: Based on information from the Kentucky Department for Natural Resources and Environmental Protection, Frankfort.

Additional information on water supply and stream pollution is given in *Acid Mine Drainage in Appalachia* (Washington, D.C.: Appalachian Regional Commission, 1969); Willard Rouse Jillson, *Pollution of Streams in Kentucky*, Kentucky Geological Survey Report, series 6, pamphlet 14 (Frankfort, 1927); R. A. Krieger and G. E. Hendrickson, *Effects of Greensburg Oilfield Brines on the Streams, Wells, and Springs of the Upper Green River Basin, Kentucky*, Kentucky Geological Survey Report, series 10, no. 2 (Lexington, 1960); John J. Musser, *Descriptions of Physical Environment and of Strip Mining Operations in Parts of Beaver Creek*

Basin, Kentucky, U.S. Geological Survey, Professional Paper no. 427-A (Washington, D.C.: G.P.O., 1963); Hayes F. Grubb and Paul D. Ryder, *Effects of Coal Mining on the Water Resources of the Tradewater River Basin, Kentucky*, U.S. Geological Survey, Water Supply Paper 1940 (Washington, D.C.: G.P.O., 1972); E. E. Jacobson, "The Water Supply Situation in Kentucky," *Journal of the American Water Works Association* 20 (1928): 854–59; Edwin A. Bell, *Summary of Hydrologic Conditions of the Louisville Area, Kentucky*, U.S. Geological Survey, Water Supply Paper 1819-C (Washington, D.C.: G.P.O., 1966); W. N. Palmquist and F. R. Hall, *Reconnaissance of Ground-water Resources in the Bluegrass Region, Kentucky*, U.S. Geological Survey, Water Supply Paper 1533 (Washington, D.C.: G.P.O., 1961); B. W. Maxwell and R. W. Devaul *Reconnaissance of Ground-water Resources in the Western Coal Field Region, Kentucky*, U.S. Geological Survey, Water Supply Paper 1599 (Washington, D.C.: G.P.O., 1962); R. W. Davis, T. W. Lambert, Arnold J. Hansen, *Water in the Economy of the Jackson Purchase Region of Kentucky*, Kentucky Geological Survey Report, series 10, special publication 20 (Lexington, 1971); R. F. Brown and T. W. Lambert, *Reconnaissance of Ground-water Resources in the Mississippian Plateau Region of Kentucky*, U.S. Geological Survey, Water Supply Paper 1603 (Washington, D.C.: G.P.O., 1963); W. E. Price, D. S. Mull, and Chabot Kilburn, *Reconnaissance of Ground-water Resources in the Eastern Coal Field Region of Kentucky*, U.S. Geological Survey, Water Supply Paper 1607 (Washington, D.C.: G.P.O., 1962).

XV. ADMINISTRATIVE ORGANIZATION

Maps 124–134 were compiled by Wilford A. Bladen.

Maps 124–127, EVOLUTION OF COUNTIES: Compiled from early maps of Kentucky in the archives of the Kentucky Historical Society, Frankfort.

For further information see Willard Rouse Jillson, *A Checklist of Early Maps of Kentucky, 1673–1825* (Frankfort: Roberts Printing Co., 1949). See also maps of Kentucky published by Thomas Cowperhwaite and Company, Philadelphia, 1850, 1851.

Map 128, AREA DEVELOPMENT DISTRICTS; **Map 129,** UNITED STATES CONGRESSIONAL DISTRICTS; **Map 130,** CIRCUIT COURT DISTRICTS; **Map 131,** STATE HOUSE OF REPRESENTATIVES DISTRICTS; and **Map 132,** STATE SENATE DISTRICTS: Based on information in *Area Development Districts Directory* (Frankfort: Office for Local Government, 1976); *Congressional District Atlas* (Washington, D.C.: G.P.O., 1975); *State Directory of Kentucky, 1976*, compiled by Mary M. Wright (Pewee Valley, Kentucky: Directories, Inc., 1976); and *1974–1975 Kentucky Local Officials Directory* (Frankfort, Kentucky: Kentucky Department of Transportation, 1974).

Map 133, WET AND DRY COUNTIES: Based on 1974 data from the Kentucky Department of Alcoholic Beverage Control, Frankfort.

Map 134, NATIONAL GUARD UNITS AND MILITARY BASES: Based on information from the office of the Adjutant General of Kentucky.

XVI. VOTING PATTERNS

Maps 135–142 were compiled by Terry L. McIntosh.

Maps 135–138, GUBERNATORIAL ELECTIONS and **Maps 139–142,** PRESIDENTIAL ELECTIONS: Based on election return data from the office of the Kentucky Secretary of State, Frankfort.

Further information on electoral behavior in Kentucky can be found in Malcolm E. Jewell, *Kentucky Votes*, 3 vols. (Lexington: University of Kentucky Press, 1963); Malcolm E. Jewell and Everett W. Cunningham, *Kentucky Politics* (Lexington: University of Kentucky Press, 1968); and *Official Primary and General Election Returns for 1975* (Frankfort: Kentucky Secretary of State, 1976).

Complete county returns for 1976 general elections and a map of voting patterns can be found in the *Courier-Journal* (Louisville), November 4, 1976.

CARTOGRAPHIC SOURCES

Cartographic literature is increasing rapidly and this selected bibliography is by no means exhaustive. It includes sources of information on map design, symbols, lettering, class-interval selection on choropleth maps, and area colors which were found useful in the preparation of this thematic atlas. There is a perceptible leaning toward current sources, and dozens of useful titles already well-known to cartographers are not mentioned. The sources cited herein should be of value to readers interested in cartographic methods.

A most useful, up-to-date, and concise guide to maps, atlases, and books on cartographic methods is Chauncy D. Harris, *Bibliography of Geography, Part 1: Introduction to General Aids* (Chicago: University of Chicago, Department of Geography, 1976), pp. 187–203, 227–28. This annotated listing gives information on general cartographic works, sources of information on maps, lists of map collections, major retrospective map catalogs and inventories, current map bibliographies, bibliographies of atlases, and works on cartographic methods. A wide coverage of research and production methods in map making and associated technologies is found in the following journals: *The American Cartographer* (The American Congress on Surveying and Mapping, Washington, D.C.); *The Canadian Cartographer* (York University, Toronto, Ontario, Canada); *The Cartographic Journal* (British Cartographic Society, London); and *Journal of the Surveying and Mapping Division* (American Society of Civil Engineers, New York).

A general discussion of current trends in thematic mapping is Joel L. Morrison, "Changing Philosophical-Technical Aspects of Thematic Cartography," *American Cartographer* 1 (1974): 5–14. Fifty-five American state and Canadian provincial atlases are evaluated by Richard W. Stephenson and Mary Galneder, "Anglo-American State and Provincial Thematic Atlases: A Survey and Bibliography," *Canadian Cartographer* 6 (1969): 15–45. Some of the state atlases published in the period 1962–1971 are reviewed by Richard W. Stephenson, "Atlases of the Western Hemisphere: A Summary Survey," *Geographical Review* 62 (1972): 92–119. State atlases published prior to 1962 were reviewed by Ena L. Yonge in "Regional Atlases: A Summary Review," *Geographical Review* 52 (1962): 429–31. A valuable discussion is provided by W. G. Dean in "The Structure of Regional Atlases: An Essay in Communications," *Canadian Cartographer* 7 (1970): 48–60.

The subject of map design has been treated by L. J. Henderson, "A Critical Look at Thematic Map Design," *Cartography* (Australian Institute of Cartographers) 9 (1976): 175–80; and M. Merriam, "Eye Noise and Map Design," *Cartographica*, Monograph no. 2 (1971), pp. 22–28. The problems of map design in atlases are examined by G. B. Lewis, "Drawing a Thematic Atlas," *Bulletin of the Society of University Cartographers* 9 (1975): 18–27. The theoretical and methodological aspects of cartography which are relevant to map design are discussed by L. Ratajski, "Cartology: Its Developed Concept," *The Polish Cartography* (Warsaw: State Cartographical Publishing House, 1976), pp. 7–23. A good discussion of cartographic perception and design is by M. Wood, "Visual Perception and Map Design," *Cartographic Journal* 5 (1968): 54–64.

The general appearance of a map is discussed by Barbara Bartz Petchenik, "A Verbal Approach to Characterizing the Look of Maps," *American Cartographer* 1 (1974): 63–71. Although somewhat dated, Arthur H. Robinson's *The Look of Maps: An Examination of Cartographic Design* (Madison: University of Wisconsin Press, 1952) remains one of the best general treatments of map design and appearance.

The basic characteristics of the communication process as it operates with maps are discussed by Arthur H. Robinson and Barbara Bartz Petchenik, "The Map as a Communication System," *Cartographic Journal* 12 (1975): 7–15; Cornelius Koeman, "The Principle of Communication in Cartography," *International Yearbook of Cartography* 11 (1971): 169–76; B. D. Dent, "Perceptual Organization and Thematic Map Communications: Some Principles for Effective Map Design with Special Emphasis on the Figure Ground Relationship," Ph.D. dissertation, Clark University, 1973; idem, "Visual Organization and Thematic Map Communication," *Annals of the Association of American Geographers* 62 (1972): 79–93; and P. C. Muehrcke, "Visual Pattern Analysis: A Look at Maps," Ph.D. dissertation, University of Michigan, 1969.

The ways in which patterns and symbols on maps are perceived by map readers are discussed by Rudolf Arnheim, "The Perception of Maps," *American Cartographer* 3, no. 1 (1976): 5–10; and G. Ekman, R. Lindman, and W. William-Olsson, "A Psychophysical Study of Cartographic Symbols," *Perceptual and Motor Skills* 13 (1961): 351–68.

A very useful study, both because of its findings and because of the indices which can be used to evaluate the communicative effectiveness of thematic maps, is Arch C. Gerlach, "Visual Impact of Thematic Maps as a Communication Medium," *International Yearbook of Cartography* 11 (1971): 194–98. Another useful examination of map communication is M. Wood, "Human Factors in Cartographic Communication," *Cartographic Journal* 9 (1972): 123–32.

The major problems in the use of areal and volumetric symbols, such as the selection of size of circles or squares with respect to the scale of the map and the range of data, are discussed by Carlton W. Cox, "Anchor Effects and the Estimation of Graduated Circles and Squares," *American Cartographer* 3 (1976): 65–74; James John Flannery, "The Relative Effectiveness of Some Common Graduated Point Symbols in the Presentation of Quantitative Data," *Canadian Cartographer* 8 (1971): 96–109; P. V. Crawford, "The Perception of Graduated Squares as Cartographic Symbols," *Cartographic Journal* 10 (1973): 87–88; Hans-Joachim Meihoefer, "The Visual Perception of the Circle in Thematic Maps: Experimental Results," *Canadian Cartographer* 10 (1973): 63–84; and idem, "The Utility of the Circle as an Effective Cartographic Symbol," *Canadian Cartographer* 6 (1969): 105–17.

Other problems relating to symbols are discussed by Michael W. Dobson, "Refining Legend Values for Proportional Circles Maps," *Canadian Cartographer* 11 (1974): 45–53; Michael W. Dobson, "Symbol-Subject Matter Relationships in Thematic Cartography," *Canadian Cartographer* 12 (1975): 52–67; and Arthur H. Robinson, "An International Standard Symbolism for Thematic Maps: Approaches and Problems," *International Yearbook of Cartography* 13 (1973): 19–26.

The special requirements of lettering for maps, consistent treatment of letter styles, the selection of suitably contrasting styles for different features, and proper arrangement of names in the total map image are discussed by Barbara Bartz, "Type Variation and the Problem of Cartographic Type Legibility," *Journal of Typographic Research* 3 (1969): 127–44; idem, "Experimental Use of the Search Task in an Analysis of Type Legibility in Cartography," *Journal of Typographic Research* 4 (1970): 147–67; P. Yoeli and J. Loon, *Map Symbols and Lettering: A Two Part Investigation* (London: European Research Office, United States Army, NTIS No. AD741834); Eduard Imhof, "Positioning Names on Maps," *American Cartographer* 2 (1975): 128–44; and D. R. Copland, "Typographic Size Control for Cartographers," *Cartographic Journal* 11 (1974): 91–93. The semiotic aspects of type faces are examined in Y. M. Pospelov, *Informative Role of Map Lettering* (Moscow: National Committee of Cartographers, 1976).

The problem of class intervals in choropleth mapping has been examined by Mark S. Monmonier, "Modifying Objective Functions and Constraints for Maximizing Visual Correspondence of Choroplethic Maps," *Canadian Cartographer* 13 (1976): 21–34; idem, "Contiguity-Biased Class-Interval Selection: A Method

for Simplifying Patterns on Statistical Maps," *Geographical Review* 62 (1972): 203–28; idem, "Class Intervals to Enhance the Visual Correlation of Choroplethic Maps," *Canadian Cartographer* 12 (1975): 161–78; and idem, "Measures of Pattern Complexity for Choroplethic Maps," *American Cartographer* 1 (1974): 159–69. Other problems in the design of choropleth maps are discussed by Gwen M. Schultz, "An Experiment in Selecting Value Scales for Statistical Distribution Maps," *Surveying and Mapping* 21 (1961): 224–30; and George F. Jenks and Fred C. Caspall, "Error on Choroplethic Maps: Definition, Measurement, Reduction," *Annals of the Association of American Geographers* 61 (1971): 217–44. A searching and comprehensive analysis of isopleths is by Mei-Ling Hsu and Arthur H. Robinson, *The Fidelity of Isopleth Maps: An Experimental Study* (Minneapolis: University of Minnesota Press, 1970).

For those interested in the use of color on thematic maps, the following works are particularly useful: D. J. Cuff, "The Magnitude Message: A Study of the Effectiveness of Color Sequences on Quantitative Maps," Ph.D. dissertation, Pennsylvania State University, 1972; idem, "Colour on Temperature Maps," *Cartographic Journal* 10 (1973): 17–21; idem, "Impending Conflict in Color Guidelines for Maps of Statistical Surfaces," *Canadian Cartographer* 11 (1974): 54–58; J. S. Keates, "The Perception of Colour in Cartography," in *Proceedings of the Cartographic Symposium* (Edinburgh, 1962), pp. 19–29; M. A. Meyer et al., "Color Statistical Mapping by the U.S. Bureau of the Census," *American Cartographer* 2 (1975): 100–107; and P. Stringer, "Colour and Base in Urban Planning Maps," *Cartographic Journal* 10 (1973): 89–94.

The technological aspects of color use in cartography are examined by Andrzej Makowski, "The Basis of Colour Technology in Cartography," in *The Polish Cartography* (Warsaw: State Cartographical Publishing House, 1976), pp. 81–100. An excellent overview of the subject is by K. Frenzel, "Map and Colour," *International Yearbook of Cartography* 7 (1967): 97–99. An interesting facet of this question has been examined by Arthur H. Robinson, "Psychological Aspects of Color in Cartography," *International Yearbook of Cartography* 7 (1967): 50–61.

A number of publications deal with the application of cartography to regional study and planning: J. J. Brandon, "Maps for Planning," *Cartographic Journal* 7 (1970): 77–80; Jan Ciesielski, *Thematic Maps for Urban Planning* (Warsaw: Institute of Geodesy and Cartography, 1976); Pape Heinz, "Urban Cartography—Town Planning," *International Yearbook of Cartography* 13 (1973): 191–98; V. T. Joukov et al., *Maps for the Long Time Forecast of Development and Location of the Economic Branches: Ways of Creating and Methods of Using* (Moscow: Moscow State University, 1976); and Tamara Brunnschweiler, "The Importance of Maps in Area Studies," Bulletin of the *Special Libraries Association, Geography and Map Division* 97 (1974): 39–41.

Useful reviews of the application of cartography to resource development problems are by A. M. Volynov, *Application of the Cartographic Method in the Study and Rational Use of Water Resources* (Moscow: National Committee of Cartographers, 1976); E. I. Gaidamaka, S. I. Nosov, and I. Y. Levitsky, *Application of Cartographic Method for Study and Rational Use of Land Resources* (Moscow: National Committee of Cartographers, 1976); S. G. Sinitsin, *Cartographic Method Application in Forest Resources Study and Protection* (Moscow: State Forestry Committee of the U.S.S.R. Council of Ministers, 1976); and Ralph W. Keifer and Michael L. Robbins, "Computer-Based Land Use Suitability Maps," *Journal of the Surveying and Mapping Division, American Society of Civil Engineers* 99 (1973): 39–62.

The foregoing paragraphs are suggestive of the kinds of cartographic literature available to students of thematic maps. The variety of sources cited should be useful to cartographers interested in thematic atlases and particularly to those who plan to undertake to produce similar volumes.

Index